WHO MADE THAT?

WHO MADE THAT?

13-Digit ISBN: 978-1-64643-215-8
10-Digit ISBN: 1-64643-215-0

This book may be ordered by mail from the publisher. Please include $5.99 for postage and handling. Please support your local bookseller first!

Books published by Cider Mill Press Book Publishers are available at special discounts for bulk purchases in the United States by corporations, institutions, and other organizations. For more information, please contact the publisher.

Cider Mill Press Book Publishers
"Where good books are ready for press"
PO Box 454
12 Spring Street
Kennebunkport, Maine 04046

Visit us online! cidermillpress.com

Typography: FreightMacro Pro, Oswald

Printed in China

All images used under official license from Shutterstock.

1 2 3 4 5 6 7 8 9 0
First Edition

WHO MADE THAT?

The Fascinating True Stories Behind the World's Greatest Inventions

TIM RAYBORN

CIDER MILL PRESS

BOOK PUBLISHERS
KENNEBUNKPORT, MAINE

TABLE OF CONTENTS

INTRODUCTION

"To invent, you need a good imagination and a pile of junk."
—Thomas Edison

Our lives are awash in inventions, both weird and wonderful. Living without some has become unthinkable. Others are just plain fun, whimsical, or pointless, albeit enjoyable. Human beings have been inventing things to make their lives easier and more entertaining almost as long as they've been around. Some of our most relied-upon inventions are literally thousands of years old, while the ideas for others—such as the camera—first appeared a long time before they could be realized. If you've ever looked at a calculator, a pair of eyeglasses, a bicycle, or a roll of toilet paper and thought, "Who made that?" then this book is for you.

In these pages, you'll learn the amazing stories behind 100 common and not-so-common objects, many of which we take for granted these days. A good number of the inventions here came into their own in the nineteenth century, when it seemed everyone wanted to be an inventor, and every inventor was trying to outdo their peers. But, as you'll see, it's curious how often several people came up with shockingly similar ideas at the same time, despite having no contact with one another. Of course, some inventors outright stole ideas from someone else and passed them off as their own (cough—Thomas Edison—cough).

Anyway, the inventions here range from the essential (the toothbrush and the personal computer) to the frivolous (silly string, the rubber duck), but each one has an interesting back story that just might transform your perspective and help you appreciate them more. This book will examine who made them and why (and interestingly, quite a few were made for other purposes first), helping you understand the human drive to invent, even when it may seem like we don't need anything else.

WHO MADE THE...
SAFETY PIN?

A little object that we could hardly do without, and yet, the safety pin can't have been around too long, can it? Well, yes and no. The safety pin as we know it is credited to Walter Hunt, a prolific and ingenious inventor who lived in the first half of the nineteenth century. In addition to the humble safety pin, he also invented (deep breath): the fountain pen, the lockstitch sewing machine, a knife sharpener, a flax spinner, a streetcar bell, a plow for removing ice, and a kind of street sweeper, among other things. Whew, that's a lot of innovation!

At the same time (1849), another inventor, Charles Rowley independently created a pin in Birmingham, England, that was similar to Hunt's iteration, but Rowley's version is no longer in use. Hunt's design is the one that stuck.

Hunt was less motivated by the thought of eternal glory when creating his miraculous little pin. Instead, it appears he was focused on paying off a $15 debt he owed to a friend (a decent sum of money in those days). After patenting the design in April 1849, he sold it to W. R. Grace and Company for $400 (calculated to be about $12,000 in modern money). He paid off his debt and pocketed a hefty $385 for himself. This may have seemed like a good financial move to Hunt at the time, but W. R. Grace would go on to mass market the safety pin and make millions on the sales. If only Hunt had known!

Was Hunt's invention really all that new? Surprisingly, no. What he called a "dress pin" was really an update of a design that had been in use since at least Roman times. These handy devices, known as fibulae, were broaches that held together cloaks, togas, and so on. Versions of these clasps were commonplace throughout Europe and beyond during the Middle Ages. Hunt's innovation was to add a little spring to help hold the pin in place.

His simple tweak succeeded far beyond anything he could have imagined, a pattern that recurred with several of his inventions. His sewing machine was innovative—but Hunt never patented it. Another inventor, Elias Howe, came up with a similar device later on. Had he managed his affairs a bit better, or recognized the utility of his inventions, Hunt could have been a very rich and respected man. But, instead of being a household name, he's barely remembered today, despite his many fine creations.

WHO MADE...
SCISSORS?

Scissors are incredibly useful tools that we could scarcely do without today. But where did they come from? Leonardo da Vinci was once credited as their originator. But, although he was a genius who invented many things, scissors were not among them. In fact, these handy tools had been in existence for far longer.

For some time, it was thought that they were invented by the Egyptians around 1500 BCE, but this has since been called into question. It seems more likely that some rudimentary shears were developed between 3,000 and 4,000 years ago. A tool consisting of two blades made of bronze and connected by a curved strip of metal has been found at the city of Emar, in what is now Syria, and dated to this period. Iron shears were first mentioned in a neo-Babylonian account from the sixth century BCE.

It was around 100 CE that the Romans invented the cross-blade, pivoting scissors that more or less provided the model for modern scissors. These Roman shears were usually made of bronze or iron, and were sometimes ornamental. Both bronze and iron blades needed to be sharpened regularly. After the first century, scissors spread both East and West via trade routes, and started being mentioned in accounts from both China and Spain.

Early forms of scissors and shears continued to be used for everything from hair trimming and fabric cutting to sheep shearing over the centuries. In 1761, a man named Robert Hinchliffe from Sheffield was the first to fashion scissors out of steel. This enabled their mass production, and to meet the growing demand, William Whiteley & Sons (also in Sheffield) began manufacturing scissors. Today, the company is the oldest operating scissor manufacturer in Europe. The production of scissors boomed during the nineteenth century, and variants in size, shape, and purpose started being manufactured. We now have a variety of types, from sewing scissors to pinking shears to the safety ones with rounded edges used in elementary schools.

An interesting scissors-related superstition: it's said you should never hand a pair of scissors directly to an individual, but should instead place them on a table or other surface and let that other person pick them up. Otherwise, you risk severing your relationship with them.

WHO MADE THE...
SPATULA?

The word *spatula* is related to the Latin word *spatha*, which was borrowed from the Greek *spathe*, which referred to either wood or a splint, but has the diminutive meaning of "sword." So, every time you use a spatula when cooking, you are, in essence, wielding a kitchen sword! Spatula in its modern sense has been used in English since about 1525, though its American and British meanings are different. In America, the word most often refers to the flat kitchen tool useful for turning over omelettes, pancakes, and the like. In Britain, it more often refers to a palette knife (such as the implement used to apply frosting to a cake), as well as tongue depressor.

The spatula has likely been flipping or probing since ancient Roman times; a legend attributes its invention to the Roman physician Galen, but as an earlier form may have existed in Egypt, it's likely Galen had nothing to do with its origin. However, there is no record of a spatula being used as a kitchen utensil in these early civilizations. Instead, the first known use was medical (again, think tongue depressors and the like).

Ancient spatulas were made of bronze and had a pointed end. They were more commonly used to mix and spread medicine, rather than for surgery. Such a large quantity of them were found at a particular site in Pompeii that some archaeologists speculate that the spatula must have performed many different functions in the ancient civilization. Some researchers believe they were used by painters to mix paint colors, for example. Perhaps spatulas started finding their way into the kitchen during this time, as it's not a large leap from these uses to realizing that it could be helpful when cooking—though hopefully they were properly cleaned before any culinary use.

It's possible, even probable, that the forerunners of the tools recovered from the Pompeii site existed long before a volcanic eruption wiped out the town in 79 CE, as there was little advancement in surgical instruments from the time of Hippocrates in the fifth century BCE until that time. Indeed, the invention of this invaluable tool probably occurred during the Bronze Age and then flourished during the Iron Age.

Some years ago, a fake article about the "inventor" of the spatula appeared online. Written as a joke, it gave the biography of one John Spaduala, an Eastern European immigrant to New York in the late nineteenth century. According to this article, Spaduala invented the spatula while working as a chef's assistant. Later, he tried to sue other companies for stealing his design, and using the name, which of course, was based on his own. There are several clues that the article is a spoof, including some references to books that don't exist, but incredibly, it has been cited as authentic and even stolen and reprinted wholesale by less-than-reputable sites. Just so you know, Spaduala didn't invent the spatula. In fact, he didn't even exist.

WHO MADE THE...
PERSONAL
COMPUTER?

We love our computers, no matter what form they take: desktops, laptops, tablets, or phones. This love has helped them become essential to much of the business conducted in the world, and many people's personal lives. Those under a certain age will find it hard to believe that there was a time in living memory when a good chunk of the population had no computers of any kind, never mind portable ones, or cell phones. The horror!

Personal computers started to come into their own in the 1980s, but computers had been around for a long time before that. The only problem was, they tended to be huge. There are many amazing graphics and jokes that talk about how a computer in the 1950s filled a whole room, and now a handheld phone has a thousand times more processing power than these antiques, and so on. And it's true. Personal computers were a distant dream in those Dark Ages, but that began to change in the 1960s.

Innovations by companies like Hewlett-Packard and IBM brought these monstrous machines down to size. In addition, NASA's need for smaller computing devices for space exploration powered many new developments, an example of how easily technology from one industry can migrate into other areas, to the benefit of everyone.

Still, not everyone thought a computer would be useful to the members of the general public. Ken Olsen, founder of the Digital Equipment Corporation (DEC), was alleged to have said that there was no reason for a "regular" person to have a computer in their home.

Despite this skepticism, IBM and other companies pushed on with the development of computers for general use. Xerox introduced a computer it called the Alto, which had some of the features of the computers we've known ever since: windows, icons, and a mouse. A few then-unknowns, like Bill Gates, Steve Wozniak, and Steve Jobs, were extremely interested in seeing where this technology could go. Companies like Tandy and Atari (yes, the video game people) created their own personal computers, and the Apple II sold astonishingly well from the late 1970s to the early 90s.

Clearly, society had developed the need for a machine that could do more than a typewriter. The public hunger for personal computers continued to grow, and as the demand increased, the supply matched it, and the price of a personal computer started to go down. Computers became not just luxury items, but valuable and indispensable tools for home users. Computers of all kinds (especially phones) are now so ubiquitous that it's seems impossible to change course—not that many would want to.

WHO MADE THE...
UMBRELLA?

An essential item for those living in the rainier parts of the world, the umbrella actually seems to have been created to ward off the sun. The word *umbrella* comes from the Latin *umbra*, meaning "shadow," but this handy device existed long before the Roman Empire. The concept of using something to shield oneself from the heat of the sun is not an especially difficult or revolutionary concept, so rudimentary umbrellas have been found all over the world. As our prehistoric ancestors probably found it more pleasant to rest and sleep under the shade of big trees, using large leaves as portable sun shelters doesn't seem that big of a leap.

Examples of early umbrellas can be found in Mesoamerica, where the Aztec culture made them out of feathers and gold. In Mesopotamia and Persia, Sargon of Akkad and King Xerxes I are both depicted in stone sculptures being sheltered by what look almost exactly like collapsible umbrellas, no doubt to reduce the effects of the punishing heat present in those regions. Umbrellas may date back as far as 4,000 years ago in India, and they were widely used at about the same time in China and Southeast Asia.

Parasols to hide from the sun were popular among the rich and fashionable of ancient Greece, though only with women; it was considered effeminate and improper for men to have them, no matter how hot and sunny it was. Their popularity carried over to Rome, and while still favored primarily by sun-shy women, it seems that some people began to realize that umbrellas could be useful in the rain, a benefit that was suitable for anyone to enjoy.

Umbrellas seem to have gone out of fashion in Europe after the decline of the Roman Empire, no doubt due in part to the chaos of that era (can you imagine a Viking carrying an umbrella in one hand and an ax in the other?), though the wealthy probably still wanted some protection from the elements, whether hot and dry or wet and cold.

Then, in the sixteenth century, umbrellas made a startling comeback, becoming fashionable in Italy. Parasols and umbrellas spread rapidly among the rich; as Italy was the center of all things chic, other nations followed along. Umbrellas were now used by both women and men, and started being depicted in art and mentioned in books with increasing frequency.

By the dawn of eighteenth century, the umbrella's use as rain protection was firmly established, and in turn, the first portable, handheld, collapsible umbrella was sold by a merchant in Paris named Jean Marius from 1710. King Louis XV was so enthralled with the device that he gave

Marius a license to sell them exclusively for five years. 1759 saw the introduction of the curved handle, a "cane" umbrella that could function as both, and could also be deployed with the push of a button. Modern umbrellas had arrived! Rainy England adopted this handy tool for widespread use by the late eighteenth century, and it's now an essential part of British life!

Depending on who you believe, steel ribs were added in either the latter eighteenth century or sometime during the nineteenth, which bolstered the structure and prevented the umbrella from being ruined by a gust of wind turning inside out (which still happens, of course, just not as frequently). Even today, new patents for improved umbrella designs are constantly being filed and approved, ensuring that some form of this handy weapon against the elements will remain in our lives for a long time.

WHO MADE THE...
LAWN MOWER?

Love them or hate them, manicured green lawns have been a significant feature of suburban homes for decades, and much longer when you think about stately homes and wealthy residences. Of course, we know that those contemporary patches of pristine grass are maintained by modern lawn mowers, but how did they manage to be well kept in the days before these mechanical creations? And when did the trusty lawn mower arrive to save the day?

The concept of the lawn bordering a home dates to the Middle Ages in Europe, when castles were surrounded by grass and fields. These areas were usually kept clear of trees and brush to remove the element of surprise from the options available to invading armies and marauding bandits—no sense in giving them cover to launch a surprise attack! But these needed to be maintained. The best way to do that? Grazing animals such as sheep, goats, and cows. Yes, the first lawn mowers were biological! Other options included using scythes and sickles—these worked, but were very labor intensive.

The word *lawn* comes from the Middle English word *laund*, meaning "a glade or pasture." The word *laund* in turn derives from the Old French *launde*, "a glade or a clearing in the woods." These clearings were often a common area near a village, where people let their animals graze.

The concept of the lawn as a ornamental element of a home really took shape sometime in the eighteenth century, though of course, the grass would have been a bit wilder and more unkempt than the immaculately manicured lawns we know today, even around the highest-caliber homes of the time.

The first mechanical lawn mower was invented and patented in 1830 by an Englishman with a Dickensian name: Edwin Beard Budding. He got the idea while observing a machine that cut cloth, and thought that the concept could be applied to cutting grass, which would be especially useful for the sports fields and gardens in Gloucestershire, where he lived. Budding's invention not only worked, it became a hit, and he successfully sold versions to the colleges at Oxford and the Zoological Gardens at Regent's Park in London. Budding and his partner John Ferrabee had the sense to license their design to other companies, rather than selling it outright, allowing them to manufacture the machines, while the two men reaped a share of the profits. They quickly sold more than 1,000 units, and other companies lined up to get in on the grass-cutting action.

There have been many mower innovations over the years (steam-powered, gas, and electric versions, for example, as well as those pulled by animals), but modern push mowers are remarkably similar to Budding's original vision, proving again that a good design never goes out of style.

WHO MADE...
SUNGLASSES?

Most would hate to be caught out on a sunny day without them, and for many, they are essential to their aesthetic. So logical, so stylish, it's hard to imagine a time when people lived without them. So, what's the story of sunglasses?

The idea is actually very old. The Inuit, who inhabit Arctic regions in Canada, Alaska, and Greenland, long ago carved bone into goggle-like devices with thin slits in them to protect their eyes against the blinding glare of the sun against snow. Evidence shows that peoples in northern Asia came up with similar inventions in order to battle the harsh light that came off the snow-covered ground.

In the twelfth century, some people in China placed smoky quartz over the eyes as a protection against bright sun and glare. It was also said that some judges would wear these in court, to hide their expressions when hearing testimonies and asking questions. Imagine modern judges doing this!

In Europe, an English optician named James Ayscough began creating tinted lenses for glasses in the mid-eighteenth century, but these were not sunglasses as we think of them. Rather, Ayscough believed that lenses with certain tints would correct various visual issues and impairments.

In 1913, British chemist Sir William Crookes invented lenses for glasses that blocked ultraviolet rays. This was a major step forward, though, at the time, few saw the advantage of these lenses, and it would be some years before the idea really took off. In the 1920s, some American movie stars began wearing them for protection against the bright Los Angeles sun, and their iconic looks made sunglasses more popular among the general populace (influencers were around well before social media!).

In 1929, Sam Foster, a plastics manufacturer, began producing celluloid for glasses, allowing sunglasses to be mass-produced and sold to people in stores. It's estimated that over 20 million pairs were sold in less than ten years. His company, Foster Grant, became renowned as a manufacturer of sunglasses, and is still going strong. Other companies, such as Bausch & Lomb and Ray-Ban were commissioned to create protective lenses for pilots and others during World War II, and the technology they developed eventually filtered into lenses produced for everyday use.

By the 1960s, sunglasses where seen everywhere, and everyone who was anyone wore them. As you can see from a scan of social media, not much has changed!

WHO MADE THE...
TOOTHBRUSH?

Everyone likes a good meal. But once it's over? It's not great to have that meal lingering in your mouth! This universally unpleasant sensation means humans have always wanted to clean their teeth, even if it was something as simple as picking at them with a twig, or chewing on some grass.

We know that the Babylonians, Egyptians, and Chinese had "chew sticks," which were essentially toothpicks. Egyptians even placed these little items in tombs so that the dead could clean their teeth in the afterlife! Chew sticks and toothpicks were commonly used in ancient Greece and Rome, and throughout Africa, where they are still popular.

Among ancient peoples, the Chinese and Indians developed a variety of wooden teeth cleaners and tongue scrapers over the centuries. The Chinese and Japanese also seem to have invented the idea of using animal hairs to actually brush the teeth. These were made from hog or horse hair and were attached to handles. (Imagine brushing your teeth with hog hair! It was probably "boar-ing.")

In the Islamic world, the *miswak* is a common tool for tooth cleaning. This small stick derives from an evergreen known as *Salvadora persica*, or the "toothbrush tree." The wood seems to have antibacterial properties and may help prevent the buildup of plaque.

Toothbrushes imported from China became popular in England by the seventeenth century, and by 1780, an entrepreneur named William Addis began mass producing toothbrushes after envisioning improvements while serving time in prison. The company he founded is still operating and sells toothbrushes worldwide, by the millions.

By the 1930s and 1940s, more improvements were on the way, the most notable the swap of natural hair for sturdier synthetic fibers, allowing for more frequent brushing (synthetic bristles also didn't hold nearly as much bacteria). Plastic handles replaced those made of wood and bone, and by the 1950s, the first electric toothbrush was invented, though the relatives of the modern versions we know didn't appear until the early 1960s.

Innovations in toothbrush tech continued, with the development of "reach" style brushes in the 1970s. These had a slight tilt in their handles and enabled the brusher to "reach" certain areas of their mouth better.

The toothbrush has become so much a part of every-day existence that in 2003 it was selected by the Lemelson-MIT Invention Index as the top invention that Americans can't live without. Of course, that was before cell phones took over the world, but it's a safe bet that having clean teeth is still at the top of most people's priority list!

WHO MADE THE...
FLUSHING TOILET?

People have always needed some place to take care of private business, but the history of their various choices would be a book in itself! This little section is confined to the flushing toilet, which are, of course, decidedly more modern.

If you're assuming that the flushing toilet must be a nineteenth-century invention (along with just about everything else we use today!), you might be surprised to learn that a kind of flushing toilet was invented in 1596. A man named Sir John Harington, none other than the godson of Queen Elizabeth I, came up with the idea. It didn't rely on a handle like a modern toilet, but instead upon a large basin positioned a floor above the seat. That basin was filled with up to seven gallons of water, waiting to be let loose into the cylinder where waste was deposited. The problem is that getting so much water up there required a lot of work, and therefore, it wasn't practical to flush after each use. Also, it made a lot of noise, and why would anyone—especially the Queen— want to announce to the world that they'd just gone? Needless to say, Harington's innovation didn't catch on.

In 1775, an English inventor Alexander Cumming formulated the idea for the "S" -shaped pipe leading to a toilet, which, when filled with water from the bowl, prevented potentially dangerous sewer gas from filling the area.

The man most commonly associated with toilets is the aptly named Thomas Crapper, who in the late nineteenth century oversaw the large-scale manufacturing and sales of flushable toilets. Crapper introduced the ballcock to the appliance, a mechanism still found in many toilet tanks today. While people might snicker at his name, believe it or not, it was entirely coincidental to his profession. The word "crap" had already been in use for centuries. But in a further twist, his toilets became known as "Crappers" after his name, not because of what one excretes while seated on them. Thus, "sitting on the crapper" actually refers to the man himself, not the business at hand. There, that's a thing you may not have known. Go forth and amaze your friends.

Toilets have seen various improvements since Crapper's time, mostly in the area of efficiency and comfort (such as heated seats for cold winter mornings!). The Japanese have elevated toilets to an art form, developing models that play pleasing sounds at the push of a button. Still, despite these improvements, the humble toilet retains what is essentially the same design it's had for over 100 years, waiting for each of us to give it our best shot.

WHO MADE...
SODA?

The world loves carbonated drinks. The amount of soda sent around the globe by the "big two" manufacturers —Coca-Cola and Pepsi—every year is staggering: 500 billion bottles of Coca-Cola alone! This obviously has all kinds of implications, from public health to the environment, but the numbers prove that soda is beyond just big business; it's woven into the fabric of our lives, a development that would have been unthinkable a century ago.

People have enjoyed soft drinks for a very long time. Beverages flavored with fruits, herbs, mint, sugar, honey, and other delicious things have been enjoyed for centuries in the Middle East and beyond. Indeed, the word *syrup* derives from the Arabic word šarāb, or "beverage."

But when we think of soda, we usually think of the carbonated stuff that bubbles up into your nose and causes you let loose embarrassing burps. Or not so embarrassing if you're a kid with trying to outdo your friends!

In 1767, English chemist Joseph Priestley discovered a method of infusing water with carbon dioxide, creating soda water. This changed the world of drinks forever. Plain soda water has since been used as a base for soft drinks and as a vehicle for hard liquors like gin and Scotch. Others improved upon Priestley's findings, including a Swedish chemist, Jöns Jacob Berzelius, who added flavors to carbonated water, including fruit, spices, and even wine. The Swiss scientist Johann Jacob Schweppe developed his own method for creating fizzy water, and the company he founded, Schweppes, is still going strong.

Soda water quickly became popular as a medicine and was prescribed for a variety of ailments, at all levels of society. In the nineteenth century, it reached well beyond the medical arena with the invention of delectable libations such as ginger beer, the ice cream soda, and that most important of all popular drinks, cola. In 1886, Dr. John S. Pemberton invented the soft drink to end all soft drinks in Atlanta, Georgia: Coca-Cola. And yes, the early form drink did include cocaine, which was widely used as medicine at the time. The original intent of the drink was as a cure-all for headaches, upset stomachs, and various other problems.

Mass manufacturing of glass bottles and later, tin cans, allowed for sodas to be sold in large quantities across the United States and beyond. In the aftermath of World War II, soda drinking was almost essential to daily life, as it still is.

WHO MADE THE...
WASHING MACHINE?

Doing laundry is rarely high on anyone's list of desirable activities, and it's a never-ending struggle to keep up. But those who have machines can be thankful the real heavy lifting is done for them. After all, who wants to go back to washboards or wringing out drenched clothes and slapping them against stones?

The handwashing of clothes was certainly a tedious business for much of civilized history. And to be honest, it's still the way the majority of the world's population cleans their clothing. Something like five billion people in the world rely on handwashing, so those who have a machine to take away that time-consuming task should be grateful!

The first washing machine design patent was issued in England in 1691, and drawings of various designs popped up over the next several decades, as inventors in Germany, Britain, and the American colonies formulated ideas about how to automate this tedious task. Most featured a large wooden barrel into which clothes, soap, and water were put. A mechanism to rotate the barrel and agitate the mixture within had to be operated by hand, but it saved considerable time compared to handwashing.

As you might expect, washing machine technology made big strides in the nineteenth century. But it was in the early twentieth century that washing machines became powered by electricity, thus removing the requirement for tedious human labor once and for all. While credited to an inventor named Alva J. Fisher, it seems that an earlier U.S. patent was issued for an electric machine, though amazingly, no one knows who it went to. Therefore, the true inventor of the electric washing machine remains unknown.

The electric washer was great for those who could afford it, though, as you might imagine, sales slowed way down during the Great Depression. It's thought that the first public laundromat was opened in Fort Worth, Texas, in 1934, but not everyone accepts this as true. The concept of a public laundry was quite old, for example, and it's not clear if this "Washateria" (as it was known) was the very first, or simply the first to really take off. Even the term itself—laundromat—dates back to the later nineteenth century.

Fully automatic washers also appeared in the 1930s, and after World War II, it increasingly became the norm to have an electric washing machine and a dryer in American homes. Europe was a bit slower to adopt the new technology, and even today, although many European homes have washing machines, many still don't have electric dryers.

WHO MADE THE...
BATTERY?

The humble battery still powers almost everything in our civilization, from cars to flashlights to cell phones. And, despite its value to contemporary civilization, the idea of the battery may be shockingly old.

Archaeological excavations in Khujut Rabu in Iraq (near Baghdad) uncovered three interesting items set together: a ceramic pot that contained a copper tube and a small iron rod. Dating them has proved difficult, but estimates place them between the third and seventh century CE. These items might not seem particularly interesting, but it's what happens when they are combined that has excited historians and archaeologists. The metal shows signs of erosion by vinegar, leading some to speculate that the liquid was used as an acidic electrolyte.

What this means is that the vinegar could have generated an electrical current, conducted through the copper and iron. What could this have been used for? The main theories are electroplating and possibly some kind of electrotherapy that resembled acupuncture. Of course, there have been many skeptics and doubters, but recent tests have shown that "wiring" up this device and using lemon juice could indeed produce electricity—up to four volts, in fact. So, did the people of the time accidentally stumble onto a simple battery that could produce small currents? Possibly.

The term *battery* was first used by none other than early America's legendary inventor, Benjamin Franklin, during his experiments with electricity. In 1800, the Italian scientist Alessandro Volta invented the first true battery, later known (rather clunkily) as the "voltaic pile." If his name sounds familiar, it should. It gave us the word *volt*, a unit of electrical measurement.

As is often the case, other inventors took the idea and ran with it. In 1859, French physicist Gaston Planté came up with the idea of the lead-acid battery that could be recharged by passing a reverse current through it. The design was revolutionary and formed the basis for many batteries to come, including the modern car battery.

Batteries continued to improve and evolve over the next century. By 1980, American physicist John Goodenough and others invented a new type, the lithium-ion battery. The technical aspects of what it does would leave many glazed over, but it was an important step in battery design. These batteries are now used for electric cars, wheelchairs, boats, and scooters, as well as for computer servers and medical equipment, as they can keep the power flowing in case of an outage.

WHO MADE THE...
CAMERA?

Cameras are something most people now think of as part of their phones—you know, those devices we use for everything other than phone calls these days. (Where would modern society be without the selfie? Probably better off, a lot of you might say!) But the camera as is has a long history, going further back than you might think.

The idea might have originated in China in the fourth century BCE. A philosopher named Mozi described and perhaps worked with the concept of the *camera obscura* (Latin for "dark room"). The concept is simple: a dark room with a small hole in one wall, through which an image is projected onto the opposite wall.

Aristotle commented on the effect, and the Byzantine mathematician Anthemius of Tralles described the effect in the sixth century. The Arab physicist Ibn al-Haytham worked with the *camera obscura* and studied optics. His *Book of Optics* (written ca. 1027) devotes much discussion to light rays and how they pass through the lens to project an image. Other medieval scientists, such as the Franciscan friar Roger Bacon (thirteenth century) studied the phenomenon, and Leonardo da Vinci was fascinated by it—his writing from 1502 is the best we have on the subject from those earlier times.

In the early nineteenth century, French inventor Joseph Nicéphore Niépce wanted to permanently capture the reversed *camera obscura* image. By the 1820s, he succeeded, taking a picture of the view from his window. It required exposing the image through the camera onto pewter coated with bitumen. The resulting image of the neighboring buildings is grainy but recognizable. Estimates for how long the exposure took vary between eight hours and several days. Not very useful for talking selfies! But it was a gigantic step forward.

His business partner, Louis-Jacques-Mandé Daguerre, came up with ways of improving the exposure and lowering the time needed to capture an image. One of the keys to this was using silver iodide, which could create much sharper images. By the 1830s and 1840s, photography was gaining considerable momentum, as inventors in several nations came up with improvements for their cameras, lifting photography from a curiosity to a necessity by the second half of the nineteenth century. And, as you know, their popularity has never waned.

WHO MADE THE...
BICYCLE?

The bicycle is fun, economical, environmentally sound, and provides good exercise. In some nations, bikes are a way of life, with literally millions of people choosing them over cars. But where did this ingenious invention come from? It would be wonderful to be able to point to an Egyptian tomb painting featuring the dead breezing into the after-life. Sadly, no such glyph exists, as far as we know.

There is an intriguing drawing of a two-wheeled machine from about 1500, attributed to an artist named Gian Gia-como Caprotti, a student of Leonardo da Vinci. Most art experts now consider this drawing to be a forgery made at a much later time, but it still has its defenders. Given how many strange contraptions Leonardo and his pupils came up with, it wouldn't be outlandish to imagine that he or one of them invented the bicycle. But the evidence doesn't seem to bear that out, sadly.

The first true bicycle prototype dates from 1817. A German man, Karl von Drais, created a two-wheeled contraption with a steering mechanism at the front. But the rider pushed it along with his feet, rather than using pedals. Still, this "velocipede" or "hobby horse" (as it was called) allowed the user to move much quicker than they could by walking.

Over the next few decades, new designs emerged, as a

good number of inventors felt that von Drais was onto something. Pedals were introduced, along with more wheels, and soon gigantic tricycles and even quadracycles were popping up. It's believed that another German, Philipp Moritz Fischer, invented the first two-wheeled pedal bike in 1853, which was lighter and easier to navigate than the heavy machines with three and four wheels. Of course, as most children remember, one had to learn the art of balancing while pedaling.

The 1860s saw more innovations, and the 1870s witnessed the rise (literally) of the high-wheel bicycle, also known as the penny-farthing. These charming, rather ridiculous-looking bicycles became all the rage for a short time, and seem to encapsulate the late nineteenth-century British inclination toward excess.

By the 1880s, bicycles were beginning to take on shapes that looked more modern. While cycling remained popular in Europe as the twentieth century dawned, it was on the decline in popularity in the United States, due to the rise of—you guessed it—the automobile as the main form of transportation. Bikes increasingly became the property of children not yet old enough to drive, and in America, that distinction largely continues—though there are no shortage of adults who cycle for exercise, recreation, and, yes, even doing errands.

WHO MADE THE...
FOOD TRUCK?

They're a pretty common sight these days: trucks that offer almost every kind of food and drink you can think of, from tacos and sushi to coffee and kombucha. They can be very simple, or upscale. While the kinds of food coming out of the food truck have become more elaborate and unusual over the last decade or so, the concept goes back a very long way. Street foods have been popular in Asian countries for centuries, for example.

In America, there is evidence of food carts in New Amsterdam (now New York) dating all the way back to the 1690s. They were an especially good idea in cramped urban areas, where there were a lot of people and few proper kitchens. In the 1860s, a rancher named Charles Goodnight (what a wonderful name!) came up with the idea of the chuck wagon. Its main purposes was to feed hungry cowboys during a cattle drive, a world far removed from the urbanized East Coast cities. These wagons were able to provide three meals a day during the long trek across the American plains. They were fitted out with cooking equipment, storage spaces, and everything else required to provide food out on the range.

By the 1890s, the food truck achieved what seemed like its ultimate form—the hot dog cart. Vendors in New Haven, Connecticut, were hawking sausages in a bun to hungry Yale students from their "dog wagons." The phenomenon became so embedded in the life of the college that the very first appearance of the term "hot dog" in print came in the *Yale Record*.

By the 1950s, people witnessed the rise of that quintessentially American tradition: the ice cream truck. With its happy tunes and slow-moving style, children in neighborhoods all across the United States happily ran alongside to snap up their favorite frozen treats in the summertime. In 1974, Raul Martinez converted an old ice cream truck and began selling Mexican food from it in Los Angeles—the birth of the taco truck!

Since then, food trucks have continued to grow in popularity and scope. In 2008, Roy Choi sparked a revolution in the LA food scene when he opened Kogi to serve Korean barbecue from a truck. Many now consider this to be the birth of the gourmet food truck.

Obviously, these trucks must follow the same health and safety regulations as restaurants, and many cities limit the number of permits that they will issue for them. Even so, estimates now place the number of food trucks in the United States at more than 24,000, a sizable fleet that does about a billion dollars of business annually. That's a far cry from those humble but handy chuck wagons out on the prairies!

WHO MADE THE...
STAPLER?

An essential office tool, the stapler has been around for a bit longer than you might think. The first prototype was actually created in the eighteenth century for France's King Louis XV. At that time, sheets of paper had to be bound together with glue, wax, or even thread, all of which were probably pretty annoying. At the king's behest, an early form of the stapler was produced. It even bore royal court insignia, indicating that it wasn't going to be made available to mere commoners. What would they do with such a dangerous and powerful technology? Perhaps start a revolution?

Wide use of the stapler would have to wait until—you guessed it—the nineteenth century. During that innovative era, use of paper increased rapidly, as more and more businesses sprung up and needed to keep records. The need for an efficient way to bundle those stacks together was ever more apparent, and filled by inventor George McGill, who received a patent for a bendable brass fastener in 1866. McGill and other inventors worked to develop the idea, and in 1879, McGill created a stapling machine that was the first commercially successful stapler, though it was large and unwieldy by today's standards.

Of course, as inventors are always trying to outdo each other, newer and better innovations on McGill's product arose over the next few decades. Now, stapler evolution might not seem like the most exciting of topics (do your best to stay away from it on a first date, for example), but this handy little machine continued to develop and be refined. In 1941, the version we are most familiar with made its appearance. And where would we be without it? It has since been responsible for countless joinings of paper, and making office jobs just a little less miserable.

WHO MADE...
CHEWING GUM?

We tend to think of gum as a wad of sugary, flavored goodness, a sticky substance we loved to chomp on as children. But we also got annoyed when we stepped in it, and grossed out when we found it under our desks at school. And we worried that if we accidentally swallowed it, it would stay in our digestive systems for seven years (this fear is completely unfounded, by the way).

Humans have been exercising their jaws on chewy substances for a lot longer than Juicy Fruit has been around. Everyone from the ancient Greeks to the Chinese, from Mexico to West Africa seems to have found a local substance that was pleasant to chew on, if not swallow. Sometimes the purpose was to cleanse one's mouth or teeth, while other rudimentary gums were mildly psychoactive or relieved pain.

In America, Native Americans in New England were known to chew on resin from spruce trees, and some colonists also took up the practice. The practice became popular enough that in the 1840s a businessman named John Curtis saw an opportunity and began to manufacture spruce tree gum, made by boiling resin and then coating it with cornstarch. This product might not seem very appealing from where we stand today, but Curtis believed in it enough to establish a factory for its manufacture. Trouble was, it didn't taste all that great and couldn't be chewed for long.

Other substances, such as paraffin wax, were substituted for the resin, but nothing stuck quite like chicle, a gum from certain Central American trees. Inventor Thomas Adams hoped that chicle could be used as a rubber substitute, but that didn't work out (just imagine trying to use chewing gum to make tires!). Also, manufacturers started to realize that making their gums taste good would make people like them more, so they started adding sugar, licorice, and other flavorings.

Of course, the man who really put chewing gum on the map was William Wrigley Jr. Wrigley first sold soap and then baking powder, and offered free packs of gum with each purchase. By 1893, he had created two brands of gum that would stand the test of time: Juicy Fruit and Wrigley's Spearmint. After their introduction, they quickly became household names. Other manufacturers popped up, and by the late 1920s, the company founded by Frank Fleer produced the first bubble gum, Dubble Bubble, and took the joys of chewing gum to a whole new level!

Packs of chewing gum still sell in the millions and show no signs of slowing down. No one has yet devised a way to make it easier to get it off the soles of your shoes, though.

WHO MADE...
BARBED WIRE?

It's a useful, if prickly invention, designed to do one thing: keep people in or out. Humans have been building fences for as long as they've been settled. They've used them for keeping animals in or out, safety, and for dealing with all of the other problems that a stationary community encounters. Barbed wire is remarkably simple, in that it requires a fraction of the material to do essentially the same thing as a fence: keep things where they're supposed to be.

The idea of barbed wire seems to be one of those concepts whose time had come. There were a flurry of patents issued in the 1860s, both in France and the United States, most of which were based on a similar idea: strips of wire with sharp barbs on them that could be stretched out and affixed to posts to effectively create a fence. The intention of these designs was to keep animals in and predators (and poachers) out. And nowhere was this more necessary than in the American West.

As ranchers took over lands and set cattle and other livestock loose, they were concerned about keeping their herds together. Wood was scarce and so large fences couldn't be built. Barbed wire seemed like the perfect solution, since it could be manufactured and sold relatively cheaply. By the later nineteenth century, there were approximately 150 companies manufacturing barbed wire to meet the demand.

Of course, some greedy ranchers tried to claim more than they rightfully owned, owing to the fact that barbed wire could be easily be stretched out overnight on public lands. The ranch wars of the time were legendary and ugly, and some historians identify the time when barbed wire became common as the beginning of the end of the "Old West."

WHO MADE...
JEANS?

They're an integral part of modern life. Jeans are worn all over the world, loved for their durability and ability to pair with almost anything while remaining fashionable. It would be hard to imagine a world without them, but they've only existed since the 1870s. Well, not quite. The fabric had existed for some time before then. "Jean" fabric was known in Italy and France going back to the sixteenth century. In fact, the fabric was known as *de Nîmes*, or "from Nîmes," and from that, of course, we get the word *denim*.

Fast-forward to the 1850s. A young man named Levi Strauss moved from Germany to New York to work with his brothers. In 1853, he moved out to San Francisco to open a wholesale business that could take advantage of the booming economy there. In the early 1870s, Strauss was contacted by Jacob Davis, a Latvian tailor in Reno, Nevada, who had a problem: the trousers he made were not tough enough for miners. The two came up with the idea of using denim and reinforcing it with copper rivets at stress points to make the clothing stronger.

Their innovation worked, and they decided to go into business manufacturing jeans for mass production. The new pants were a previously unseen blend of rugged and comfortable, and over the coming years, the design was tinkered with until the famed blue tint and 501 design arrived in the 1890s.

Jeans have always been popular with workers, but it was during the 1950s that they came into their own as a symbol of cool. James Dean wore them in the film *Rebel Without a Cause*, and this began to fire the imaginations of would-be rebellious youths everywhere. The 1960s beckoned and wearing jeans became a symbol of youth, change, and a refusal of social conventions that had grown stale. By the 1970s, that fire had spread to the extent that jeans were normal clothing worn by pretty much everyone. The fashion for jean-wearing spread around the world, and even cultures that had little to do with America seemed eager to adopt them.

In the 1980s, jeans went high fashion, with expensive "designer" jeans becoming all the rage. Stonewashed jeans were another popular trend during the '80s. Still, despite all of these developments, and companies offering jeans in every color, the original Levi's blue jeans are still the top choice of consumers.

WHO MADE THE...
TELEPHONE?

We use our phones these days for just about everything except ... talking. It's amazing that what we call phones are actually just computers that let us carry around our entire lives in our pockets. It's astonishing that just twenty years ago, portable phones weren't all that common. Most people of a certain age now probably wonder how they ever got along without theirs, and those who are young enough to never have known an age before cell phones can't even begin to process the prospect of not having one.

But listen up, youngins: amazingly, there was a time when all a phone did was let you to talk to someone. It was attached to a wall, and you had no idea who was calling before you picked it up and said "Hello." Yes, it was terrifying! And it was the way of things for over a century.

But where did this device originate?

Most people probably know that Scottish inventor Alexander Graham Bell invented the telephone, but there's a little more to it than that. The ability to communicate over long distances was something that people had desired for some time. As early as the seventeenth century, British scientist Robert Hooke experimented with ways of talking over a vibrating wire. The "tin can" phone, where two cans are connected by a piece of long string, allowing people stationed on each end to talk to each other, is basically just a realization of Hooke's idea.

In the early nineteenth century, the telegraph allowed for electrical signals to be sent across long distances. Instead of speaking, patterns were tapped out, the most famous system of which was devised by Samuel Morse in 1837. Known as Morse code, it's an elegant system that's still in use.

From the 1850s to the 1870s, a number of inventors laid claim to creating the telephone, including Bell, Antonio Meucci, Elisha Gray, and Thomas Edison. Each invention was a variation on the same idea, so it's difficult to say exactly who "invented" the telephone. Bell was the first to patent a telephone, but debate about the issue continued right into the early 2000s, with both the American and Canadian governments weighing in. The United States recognized Meucci's contributions, while Canada overwhelmingly endorsed Bell, the Canadian immigrant. The controversy still hasn't completely gone away.

In any case, the phone was here to stay by the 1870s, as businesses began to see their utility. By the 1880s, they had found their way into homes. And although no one was yet taking selfies with them, the phone became established as an unquestioned piece of everyday life.

WHO MADE THE...

NAIL CLIPPER?

Since humans have been around, they've had to contend with nails that grow involuntarily. And for an untold number of years, the only way to keep them short enough that they didn't get in the way was to bite them off—and yes, this was probably true for both hands and feet! Fingernail-biting remains a habit that a lot of people indulge in, even if they're embarrassed to admit it. But think about it: it's a practice with millennia of tradition on its side!

With the rise of civilization, alternatives appeared, primarily trimming nails with small knives. Of course, one had to be very careful not to cut up their fingers and toes, and undoubtedly, a lot of accidents happened, creating a whole new set of issues. Over time, some superstitions developed around nail trimming, including this odd account from an 1889 *Boston Globe* article:

> "It is unlucky to cut the finger nails on Friday, Saturday or Sunday. If you cut them on Friday you are playing into the devil's hand; on Saturday, you are inviting disappointment, and on Sunday, you will have bad luck all week. There are people who suffer all sorts of gloomy forebodings if they absentmindedly trim away a bit of nail on any of these days and who will suffer all the inconvenience of overgrown fingernails sooner than cut them after Thursday."

Well, that settles that! Fortunately, fingernail clippers had arrived by the time this article appeared, making it easy for people to stick to the Monday-to-Thursday window for trimming! The first patent for nail clippers was issued in the United States in 1875, given to a Massachusetts man with the wonderful name of Valentine Fogerty. A better, more refined version seems to have been created by Eugene Heim and Celestin Matz in 1881, and several more designs followed over the next few decades. Like so many of the inventions in this book, once one design caught on, other inventors looked for ways to improve on it, i.e., cash in.

The clippers we now know and love had begun to take shape by the beginning of the twentieth century. The most familiar model appeared in 1947, when William E. Bassett made some more modifications and created the version still in wide use currently. Today, The W. E. Bassett Company produces nail clippers and many other products that are sold all over the world.

WHO MADE THE...
PAPER CLIP?

As long as there has been paper, there has been a desire to bind sheets of it together. And while the stapler (see page 58) is one option, sometimes you don't want something quite so permanent, or to sully your paper with holes. Enter the paper clip. It seems so obvious an invention that it should date back centuries, but paper clips are a relatively recent arrival on the scene.

The first patent for a paper clip–like object was granted to the American Samuel B. Fay in 1867. Curiously, the implement was not intended to hold sheets of paper together, but rather, to clip paper (such as tickets) to fabric. The most obvious use would be at a coat check or cleaner. The patent did allow, however, that sheets of paper could also be clipped together with it, which was a mighty decent concession! As usual, various other designs followed over the next few decades, but the one that became the paper clip's avatar, i.e., the curved wire, was never actually patented, which seems almost ridiculous, given how it took over the world!

That familiar design is known as the Gem paper clip. The earliest surviving advertisements for it date to 1893, as a product of the British Gem Manufacturing Company. The Gem style quickly became the most popular form of clip, simply because it works so well.

Of course, this didn't prevent other styles and shapes from being invented and marketed, some of which are still used in offices today. But the handy curved clip is the one overwhelmingly favored by professionals, schools, and in the home. It's just a shame that we don't know who invented it!

WHO MADE THE...
TEDDY BEAR?

Most probably had one or more of these lovable companions in their childhood, and they seem such a staple of youth that it's hard to imagine a time when they didn't exist. It's true that other toys, dolls, and stuffed animals have served as children's companions for a very long time, possibly even into prehistory. But this plush, cuddly bear can trace its origins to a very specific point in time.

U.S. President Theodore Roosevelt was on a hunting trip near Onward, Mississippi, on November 14, 1902. The party hadn't found any bears, so his assistants chased down a black bear, captured it, and tied it to a tree, suggesting that the president shoot it. Roosevelt refused, saying that this was unsportsmanlike (though he did suggest that they put it out of its misery).

News of the story spread quickly, and political cartoonist Clifford Berryman lampooned it for the *Washington Post* on November 16, 1902. A Russian-born candy shop owner in New York, Morris Michtom, saw the cartoon and had an idea. His wife already made stuffed animals of various kinds, so he asked her to create a plush bear cub. She did and he sent it to Roosevelt, calling it "Teddy's Bear." The president was pleased and gave Michtom permission to use his name on it as a product. So another bear went up in the window of the shop, and soon people were inquiring as to how to buy them. A new industry had been born!

Sales were so successful that Michtom set up the Ideal Novelty and Toy Company to manufacture and sell "Teddy bears." The company became a mainstay producer of toy and games in America, lasting until 1997, when it was absorbed by other corporations.

At the same time (because there always seems to be something in the air surrounding those innovations that become embedded in our lives), a designer in Germany named Richard Steiff created a stuffed bear that was shown at the Leipzig Toy Fair in March 1903. A German buyer named Hermann Berg (brother of famed modern composer Alban Berg) ordered a set to be delivered to a company in New York, but either the order was never fulfilled, or the bears did not survive the ravages of time. In any case, it was Michtom's bear that captured the American imagination, and in time, the "teddy bear" became the best friend of children all over the world.

WHO MADE...
WINDSHIELD WIPERS?

Obviously, windshield wipers came about after the automobile, but they weren't something that the earliest car models featured, which must have made driving in the rain much more difficult and perilous—if not impossible.

A man named George J. Capewell from Connecticut is on record as receiving a patent for a type of wiper in 1896, and other designs soon followed. The most famous of these was one devised in 1903 by Mary Anderson, a real estate developer and inventor. As the story goes, Anderson was riding in a New York street car during a snow storm and saw that the driver was having a very difficult time keeping the windshield clear, either leaning out of the car to clean off the glass or stopping the whole train so that he could get out and wipe it off when conditions got really bad.

Anderson saw an opportunity: a windshield wiper that could be operated from inside the vehicle, saving time and effort, and making travel through inclement weather a lot safer. She is often credited with "inventing" the windshield wiper, but there was also a wave of inventions and patents from people working off a similar premise in both the United States and Europe from the 1890s through the 1910s.

This glut of options prevented Anderson's design from catching on initially, as more than one company rejected it as not having enough practical value or marketability. Personal automobiles were still pretty rare in those days, so no one had the foresight to see what might be coming. Later, when cars were the norm and the industry was booming, many manufacturers went back to Anderson's ideas and designs and saw value in them. In 1922, Cadillac became the first company to make windshield wipers standard on all its car models. After that, the idea really took off, and drivers appreciated not having to stop, get out of their vehicles, and clean their windshields every few minutes!

Mary Anderson lived until 1953, seeing her practical invention widely adopted, so she had the last laugh!

WHO MADE THE...
ICE CREAM CONE?

Ice cream is one of the signature tastes of summer, and for many people, that love lasts all year round! And having it in a cone is even better. You may think that something as simple as the ice cream cone would have an uncomplicated origin—but you would be wrong. The number of competing claims about who invented the cone, and when they did, is kind of shocking.

As early as the mid-eighteenth century, it seems that wafers were being rolled and served with (or stuck onto) ice cream, but whether they were used in the same way as the classic cone is unclear. In Charles Elm Francatelli's book, *The Modern Cook*, published in London in 1846, he mentions filling little "cornets" with ice cream and using them as a garnish for larger frozen desserts. It's clearly related, but are these trimmings actually ice cream cones?

By the late nineteenth century, several English cookbooks had recipes for similar treats. They were usually cornets filled with cream, fruit, and ice cream, and served as a side at lunch or dinner. In 1894, Charles Ranhofer, a chef at Delmonico's Restaurant in New York published a cookbook called *The Epicurean*. In it, he offered a recipe for waffles, rolled into cones and filled with cream. Again, close, but still not quite there ...

The ice cream cone we know and love seems to have developed among Italian immigrants in London during the nineteenth century. Several of them started businesses London's Little Italy. Ice cream was a favorite treat for young and old alike, though most often it seems to have been sold in glasses, which were returned to the counters, washed, and reused.

Antonio Valvona, an Italian immigrant in Manchester, England, applied for a patent in 1902 for a special machine that would "make cups or dishes of any preferred design from dough or paste in a fluid state that is preferably composed of the same materials as are employed in the manufacture of biscuits, and when baked the said cups or dishes may be filled with ice-cream, which can then be sold by the venders of ice-cream ..."

Around the same time, the concept of the ice cream cone began to take hold in New York. In 1903, Italo Marchiony applied for a patent for a device that molded dough into cones to hold ice cream. Several other inventors and vendors seemed to have similar ideas around this time. One story claims that at the 1904 World's Fair in St. Louis, Ernest A. Hamwi was selling waffle pastries next to an ice cream vendor. When the ice cream seller ran out of cups, Hamwi made a proposition: he would fold his waffles into cones that could hold ice cream, and thus a tradition was born!

As you can see, there are many different stories that point to the origins of the humble ice cream cone (including many more not mentioned here). It is hard to know which one has the strongest claim, but there is no doubt that Italian food vendors in England and New York were largely responsible for the existence of this sweet, crunchy treat.

WHO MADE THE...
PAPER TOWEL?

Paper towels have been the hero during an infinite number of kitchen spills, slops, and disasters, making cleanup easier than it would be with a cloth. Of course, there are environmental issues to consider when using them, causing many manufacturers to make paper towels from recycled paper in recent years.

The paper towel may have been invented completely by accident. In 1879, Clarence and Irvin Scott, brothers from Philadelphia, started a paper manufacturing company. The company did well selling facial tissue and toilet paper, but in 1907, Arthur Scott (son of Irvin) faced a big problem. An entire railroad car of paper sent to his manufacturing plant had been rolled too thick to be used as toilet paper.

Faced with having to send it back or not use it at all, Arthur remembered that a Philadelphia teacher had come up with a unique answer to a persistent health concern: children spreading germs at school. Instead of having kids wipe their hands on a cloth towel after washing them, she gave them a small piece of paper instead, so that they could dispose of it afterward.

Scott had an idea; he perforated the too-thick paper into small sheets, called them "Sani-Towels," and offered them for sale to hotels, train stations, and restaurants as disposable towels for their washrooms. These businesses were very eager to buy his invention and the idea caught on quickly.

While this accidental invention was well received by businesses, the idea of Sani-Towels for home use would take same time. The Scott Company introduced "paper towels" for the kitchen in 1931, but they weren't an immediate success. Many housewives were skeptical. What good were they, when dishes still needed to be dried with cloth towels? It took some time for the idea that these towels would be great for cleaning up without the hassle of doing laundry afterward to catch on.

But, of course, the idea did take off eventually, and paper towels became an essential part of any kitchen. As far as home paper products go, paper towels are now only outsold by toilet paper, making Scott's brilliant gamble one of history's best examples of taking lemons and making lemonade!

WHO MADE THE...
COFFEE MAKER?

Coffee is the drink that fuels the world, for better or worse. Where would we be without coffee? Not a question most would want to consider, much less answer. People have used many methods to brew coffee over the centuries, but coffee makers are a relatively new invention—and one that makes mornings much easier for many.

The origins of the coffee maker depend on how you define "coffee maker." There have been a number of handy devices over the centuries, including an early form of the French press that dates to 1806. But if we're talking about machines to make coffee, those come a bit later. The idea of the percolator was first introduced by an American inventor named James Nason in 1865. Later on, a more sophisticated version was patented by Hanson Goodrich. But in both cases the coffee produced didn't taste all that great, because the percolator overcooked the grounds, lending the brew an extremely bitter quality.

But people's thirst for coffee demanded that development push on. The first espresso machine was built by the Italian inventor Angelo Moriondo in 1884, but he failed to capitalize on it, and it never really took off. It wasn't until 1901 that a man from Milan, Luigi Bezzera, improved upon Moriondo's design and created an espresso machine that was far more workable. His patent was bought and the machine was displayed at the 1906 Milan fair; true Italian espresso was born!

Meanwhile, in 1908, German entrepreneur Melitta Bentz came up with a crucial concept: the filter. This enabled better-tasting and far more enjoyable coffee by filtering out the grit. Over the next several decades, there were seemingly endless innovations in coffee machine technology, from the modern French press to the automatic drip pot, created by the German inventor Gottlob Widmann in 1954.

By the early 1970s, the automatic drip machine had become the most popular form of coffee maker in the United States, supplanting the percolator. The early machines were made by Mr. Coffee, but soon, other manufacturers, such as Cuisinart, Braun, and Hamilton Beach offered their own versions of what was basically a home version of a restaurant coffee maker. As time went on, home versions of espresso machines and a variety of other inventions, such as machines by Keurig and the Verismo from Starbucks, came along. All of these coffee creations will ensure that no one will be deprived of their caffeine fix for a long time to come!

WHO MADE THE...
TOASTER?

Toast is an essential piece of the breakfast table. Or something that's eaten on the run with coffee on those late-for-work mornings. While humans have been toasting bread over fires for a long time (usually burning one side or the other badly), the modern toaster only appeared on the scene relatively recently.

The first electric toaster was created by a Scottish inventor, Alan MacMasters, in 1893. Of course, this was not the pop-up type that we know and love today. For the next decade or so, these rudimentary toasters only toasted the bread one side at a time, and the machine needed to be turned off after the toasting was done— quite inconvenient for those with a tendency to be tardy in the morning.

Another issue was that the wires used to conduct the high temperatures required to brown bread tended to break easily—imagine how annoying having your toaster break in the middle of the process would be. Now imagine it happening once every week or two. The problem was solved in 1905 by an American metallurgist named Albert Marsh, who used an alloy of chromium and nickel to make the wires strong enough to withstand numerous instances of heating and cooling.

That's all well and good, but what about the pop-up toaster capable of toasting bread on both sides? In 1913, American inventor Lloyd Groff Copeman came up with a machine that turned the bread, removing that tedious (and potentially finger-endangering) task from the user. In 1921, inventor Charles Strite created a toaster that—yay!—popped up the bread when toasting was finished. In 1926, Waters Genter Company made some tweaks to Strite's design and put out the Model 1-A-1 Toastmaster, the first pop-up toaster for the home. The Toastmaster also had a timer so that the bread wouldn't overcook. This new toaster was a hit, and countless redesigns and tweaks have happened since, reimaginings that include everything from mini toaster ovens to "smart toasters" that have computers in them.

These newest iterations allow for perfect toasting of whatever you put in, from slices of bread to croissants and bagels. With some, you can toast different types of bread at the same time—a long way from holding hunks of bread over an open flame!

WHO MADE THE...
BAND-AID?

Band-Aid is actually a trademarked name owned by Johnson & Johnson. But, as these sticky little items have been a panacea for countless people dealing with minor injuries since they were introduced in the early twentieth century, that term has become generic, the stand-in for any adhesive bandage.

The Band-Aid was invented by a Johnson & Johnson employee, Earle Dickson, in 1920. Dickson, seeing that his wife seemed a bit accident prone in the kitchen, wanted to find an easy way for her to dress her wounds. So, he combined two products Johnson & Johnson already produced, gauze and adhesive tape, and invented the first adhesive bandages. Dickson covered the bandage with some crinoline fabric to prevent it from sticking to itself, and a prototype of the Band-Aid was born!

Realizing that he was onto something that could be helpful beyond his own house, Dickson showed his idea to his supervisor, who in turn introduced him to the company's president, James Wood Johnson. Johnson was intrigued, and agreed to produce the bandages. Initially, the company handmade the bandages, and they struggled to catch on with consumers. But, convinced they had a strong idea, the company started using machines to make the bandages in 1924. This gave them a much more uniform shape and size, and the public responded as the company had initially expected.

Band-Aids have since helped ease the effects of an infinite number of boo-boos and small scrapes. World War II saw a huge increase in production, as millions of Band-Aids were shipped to American troops in both theaters. They proved invaluable for the endless scratches and scrapes that soldiers endured almost daily.

As for Dickson, he had a long and successful career with the company, eventually working his way up to vice president. He retired in 1957, and by then Band-Aids were a staple of the American medicine cabinet. His attempt to help his wife had revolutionized the way the world treated simple wounds, and likely prevented a lot of infections and hospital visits as well.

WHO MADE THE...
EGG CARTON?

Unlike many inventions in this book, the egg carton has a definitive starting point and a simple, straightforward evolution. A journalist and publisher in British Columbia named Joseph Coyle happened to overhear an argument between a hotel owner and a farmer. The proprietor of the hotel was complaining about how many of the eggs he'd purchased arrived broken. The farmer didn't really have a solution; eggs were fragile, you know?

So Coyle had an idea. Would it be possible to find a way to transport eggs that kept them safe? He experimented with newspaper (since he had plenty on hand) and soon came up with a simple, ingenious solution: placing an egg in a little container shaped exactly like an egg, so that it wouldn't move around during transport? Perfect!

At first, Coyle made his carriers by hand, but once word got around, the product really began to take off. So, Coyle patented the concept and turned to automation to meet the demand, establishing a factory in Vancouver.

Two more inventors followed in Coyle's footsteps, adding their own improvements. In 1921, Morris Koppelman created a container that was more like the modern egg carton, and in 1931, Francis H. Sherman patented a design made out of pulped paper. Sherman's design is the one you'll find in grocery stores all over the world today.

Over the years, there have been other versions of the egg carton, including some made out of plastic, but on the whole, the humble egg carton is still pretty similar to the initial ones Joseph Coyle created to soothe an angry hotelier and his broken eggs. Who knew a small argument could change the world?

WHO MADE THE...

TRAFFIC LIGHT?

There's nothing more annoying that having to wait at a long traffic light. Unless it's just barely missing the green. Traffic lights are essential to any place that has even a modicum of automobile traffic, and love them or hate them, the world couldn't do without them.

Interestingly, traffic lights first appeared before the car even existed.

The first prototype of a traffic light was a gas-lit lamp installed outside the Houses of Parliament in London in December 1868, to control the flow of horse-drawn carriages. J. P. Knight, a railway engineer, came up with the idea by adapting the gaslit signals used for trains. The traffic light used green for traffic to go and red to signal for it to stop, but unlike modern lights, it had to be operated by a police officer (police had already been directing traffic in the area for many decades).

The light worked just fine ... until, less than one month into its run, one of the lights exploded due to a leak in the gas line. It injured the police officer on duty, and that was pretty much the end of that experiment.

Fast-forward to the age of the automobile. The need for some way to control city traffic became more apparent with each passing year. Various stop-and-go systems were tried out in a number of cities, again usually operated by police officers. Sometimes they used colors; sometimes they had signs that said "stop," and "go" or "move."

In 1912, a police officer (the fittingly named Lester Wire) in Salt Lake City, Utah, came up with the idea of an electric light that used the familiar red and green colors. More lighting innovations came to Cleveland in 1914, and then to Detroit, where another policeman, William Potts, devised the first three-color traffic light that worked at a four-way intersection. Having this invention in the city that was the home of automobile production lent it extra credibility, and cities around the United States began to see the value in using this simple, effective tool to control traffic and prevent accidents.

In 1923, African American inventor Garrett Morgan was granted a patent for his version of the electric, three-light machine. After that, traffic lights took off, so to speak, and were adopted around the world over the next several decades. Traffic lights finally returned to London in 1926, being installed at Piccadilly Circus, which isn't all that far from the Houses of Parliament. Fortunately, this time, nothing blew up.

WHO MADE...
SLICED BREAD?

You've probably heard the phrase "the greatest thing since sliced bread," a statement that makes it sound like this revolutionary change came about at some point during the Enlightenment. But commercially available sliced bread isn't all that old, not even 100 years as of writing. Now, of course, people have been slicing their bread for as long as they've had knives, but commercial sliced bread only came about in 1928.

The inventor of the machine that could accomplish this Herculean task, Otto Rohwedder, had a bit of a struggle getting his creation out to the world. He'd designed a prototype way back in 1912, but a fire destroyed not only it, but also his blueprints, meaning he had to start all over again. Fortunately, Rohwedder wasn't deterred, and by 1928, he'd made a commercially viable model. The Chillicothe Baking Company, based in Chillicothe, Missouri, sold its first loaves of pre-sliced bread in July 1928.

Not all bakeries were impressed; many thought slicing the bread beforehand would cause it to go stale sooner. But the momentum of Rohwedder's idea could not be stopped, and a local newspaper promised consumers that they'd experience a "thrill of pleasure" at seeing bread cut into perfect, even slices. The paper went on to say that this sliced-up wonder was sure to "receive a hearty and permanent welcome."

As we know, that article ended up being spot on. Sliced bread caught on, and became ever more popular over the next decade. Of course, the really big question is: what was the greatest thing before sliced bread?

Fun fact: sliced bread was briefly banned by the U.S. government in 1943, from January to March. This move was not due to worries about pre-cut loaves being a nefarious plot of the Axis Powers. It was simply a practical decision, in order to save resources. It was thought that since the bread required a thicker wrapper to keep it fresh, it was using up materials quicker. There was something of a public outcry over the decision, and the ban was reversed two months later.

In any case, sliced bread came roaring back, and thrived in the post-war years. It's still the primary way that most commercial brands are sold at grocery stores, and it's still the standard by which all inventions are measured.

WHO MADE...
SCOTCH TAPE?

This see-through tape has been a lifesaver for many household needs and crafts, to say nothing of the infinite number of Christmas and birthday presents it has secured in wrapping paper over the last 90 years! According to the Scotch brand's website: "In 1930, a young banjo player-turned-3M engineer named Richard Drew invented the first transparent tape." Bet you weren't expecting to read that! So, who was this young banjo player, and why did he invent such a useful gizmo?

Well, Drew wasn't a full-time banjo player. He worked for 3M in St. Paul, Minnesota, a company that made nothing but sandpaper in the 1920s. At an auto body shop one day, Drew happened to notice that the painters were finding it difficult to get a clean line of division between two colors of paint. The problem got him thinking, and he eventually came up with an idea for what would become masking tape. It was an ingenious idea that has since become a standard way of dividing areas (on walls, cars, wherever) that need to be painted, or not painted.

But one day, while testing out Drew's tape, the painter at the body shop was unhappy with how sticky, or rather, un-sticky, it was. Frustrated, he told Drew: "Take this tape back to those Scotch bosses of yours and tell them to put more adhesive on it!" In those days "Scotch" was a pejorative word meaning "stingy" or "parsimonious," which may or may not have had bigoted undertones. But Drew decided to go with it and call his line of adhesives "Scotch Tape."

And in 1930, Drew did something that gift-givers have been grateful for ever since: he invented a new kind of Scotch Tape that used cellophane, which are transparent sheets made of regenerated cellulose. The design was further improved in 1961 with the introduction of "magic tape." This is the now-famous tape that appears opaque on the roll, but becomes virtually invisible when pressed on surfaces. It can also be written on, unlike glossier tapes.

In 1939, 3M produced its iconic tape dispenser, the "snail," an innovation that made using Drew's line of tapes even easier. Leaning into *Scotch* as a name to be proud of, 3M adopted the famed tartan design that still adorns its tape products in 1945, and the brand "Scotch" was soon applied to other products, such as Scotchgard, Scotch-Brite, and even cassette and video tapes.

Today, Scotch Tape has become a brand so successful that it has entered into the realm of the generic, with people using it even if they're referring to other kinds of tapes. The young banjo player from St. Paul would be thrilled to know just how far his invention has spread, and just how much happiness it has helped spread!

WHO MADE THE...
BALLPOINT PEN?

This little implement has been one of the best-selling and most useful devices ever made. Prior to its invention, people were forced to write with fountain pens, which, while elegant, can also be messy, unreliable, and leave ink stains everywhere. The ballpoint does away with such hassles, keeping the ink firmly inside the pen and away from your fingers, clothing, and furniture.

Although it didn't really see the light of day until the 1930s, a prototype for the ballpoint pen existed as far back as 1888, when a tanner named John J. Loud took out a patent for a pen that relied on a ball and could be used to write on tough surfaces like leather and wood. It worked, sort of, but was too rough to be used on paper, and the idea never caught on. Eventually, Loud's patent expired, and the idea was largely forgotten. And people continued to get ink all over the place.

The charge of finding a solution fell to László Bíró, a Hungarian newspaper editor. Taking a cue from newsprint ink, which tended to dry faster than the ink used in fountain pens, Bíró devised an idea for a pen that would use the same kind of ink, and a ball-socket mechanism to keep the ink in place. With the help of his brother, György, László created a pen that would be durable, wouldn't leak, and could write effectively on paper.

After obtaining patents for the pen in 1938, the brothers fled from the threat of World War II to Argentina, and opened a pen manufacturing factory. Their pen proved to be a big hit, and the company was soon selling to prestigious clients like the RAF in Britain. Indeed, these pens are still known as "biros" in the UK. Other pen companies started popping up, tweaking the design to obtain new patents.

In the 1950s, Italian manufacturer Marcel Bich got in on the game. In 1953, he licensed a design from Bíró, and sold them under his own name, shortened to "Bic" to prevent any unfortunate mispronunciations! Today, Bic ballpoint pens are known throughout the world, and have been a mainstay of offices, schools, and homes ever since.

WHO MADE THE...

CHOCOLATE CHIP COOKIE?

A classic dessert, enjoyable any time of day (anyone who says chocolate chip cookies aren't appropriate for breakfast is clueless). But this standard hasn't been with us for all that long; just a little over eighty years as of writing. And while some inventions are shrouded by the mists of time, we know exactly who invented this delicacy, and where.

American chef Ruth Graves Wakefield ran the Toll House Inn and Restaurant in Whitman, Massachusetts. There are several apocryphal tales about how this little bit of culinary history came about. One story says that she was going to use nuts in her cookies but had run out, and decided to substitute chocolate chips instead. Another tale claims that she had assumed the chips would melt completely and produce chocolate cookies. An even sillier version says that it was total an accident, with the vibration of a machine causing the chips to accidentally fall into the cookie dough mix.

The truth is less colorful but proves that Wakefield was forward-looking and always up for experimenting. She said that she simply wanted to impress the guests with something new, and came up with the idea. "I was trying to give them something different. So I came up with Toll House cookie." She added bits of chopped-up, semisweet chocolate from a bar into the mix, and a masterpiece was born.

The cookies were an instant hit, and in 1939, Wakefield decided to give Nestlé the rights to use the recipe and the Toll House name. This platform was enough to create an instant classic beloved the world over. Wakefield was reportedly told she would be paid—wait for it—one dollar for the rights. But apparently, she never received even that. However, she did receive free chocolate for life from Nestlé, so maybe it wasn't such a bad deal after all. She was also paid to consult with them on future projects. The chocolate chip cookie became a standard of kitchens everywhere, beloved by generations of children and adults.

The Toll House Inn and Restaurant is long gone (it burned down in 1984), but Wakefield's creation lives on, more popular than ever!

WHO MADE...
DUCT TAPE?

There's nothing that needs fixing that can't be fixed by duct tape ... allegedly. Or is it "duck tape"? The answer is: it's actually both. What we now know as duct tape was originally commonly called "duck tape," owing to its water-resistant capabilities and duck mesh cotton, which provided its legendary strength. Versions of this duck tape date back to the early twentieth century, where it was used for everything from strengthening clothing to wrapping steel cables and reinforcing boots.

The silver tape we know and love today was an innovation of World War II. The idea came from Vesta Stoudt, a factory worker whose two sons were in the U.S. Navy. Based on their reports from the front, Vesta became worried about the seals on ammunition boxes, and wanted to create a more efficient and safer method to open and close them. In 1943, she wrote to President Franklin D. Roosevelt about the problem—as well as a potential solution that she had come up with.

The president was interested in Stoudt's idea and sent it over to the War Production Board, which then contacted Johnson & Johnson. Working with Stoudt's concept, the company developed a strong new tape reinforced with duck mesh cloth and coated in waterproof polyethylene. It could be ripped by hand, rather than needing to be cut with scissors. It was applied and removed easily, but was strong enough to supply a really good bond. Unlike the modern version of the tape, the Johnson & Johnson version was "army colored"—olive green—to match much of the equipment it would be used on. The roll-out (no pun intended ... well, maybe a little) was a success, and the tape proved very useful for the remainder of the war.

In the 1950s, the tape was commonly used in various industrial settings, especially to wrap air ducts, which seems to have provided "duct tape" moniker. This was also the time that it began to appear in its classic silver color, so that it could be used on metal pipes and vents without standing out.

In 1971, American businessman Jack Kahl acquired the rights to the sticky substance and rebranded again as "Duck Tape," playing off of the old name and making it "official." The new label even included a cartoon duck logo to reinforce the idea.

WHO MADE...
SPRAY PAINT?

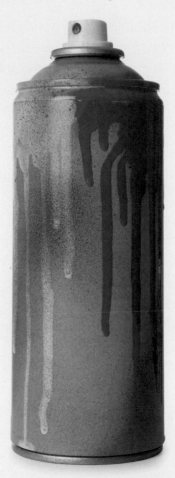

Beloved by graffiti artists everywhere (and hated by many for the result of that passion), spray paint allows paint to be quickly applied to surfaces without any need for a brush or getting your hands dirty. Quite useful in industrial settings and for painting cars, it only came about fairly recently.

Spray paint is a combination of two concepts: paint and the aerosol can. Aerosol has been toyed with since the late eighteenth century, and several inventors conceived of using a pressurized aerosol container to expel a substance during the nineteenth century. In 1927, Norwegian engineer Erik Rotheim invented and patented a can with a valve that could expel the contents. During World War II, further innovations came along.

From there, it wasn't long before spray paint made its appearance. In 1949, Robert H. Abplanalp came up with the idea for a crimp on the valve, using the pressure of inert gas to expel a liquid. This was also the beginning, for better or worse, of modern insecticides. Initially, these bug sprays were primarily used by soldiers to battle the disease-bearing insects that tend to collect in the world's tropical climes.

That same year, paint salesman Edward Seymour of Illinois invented the first can that could spray paint, on the suggestion of his wife, Bonnie. Seymour had been looking for a way to quickly and effectively paint radiators with an aluminium-based paint, and this seemed like the perfect solution. It worked like a charm, and Seymour knew he was onto something big. He started a company, Seymour of Sycamore, Inc., to manufacture spray paints and their cans, and the whole thing took off.

Graffiti artists started to make names for themselves in the late 1960s, when protest against the establishment was all the rage. Spray paint seemed like the perfect way to get bold messages across, and while many people rightly loathe the mess that graffiti can make on buildings and walls, some have elevated graffiti to a high art in the decades since. It's probably not what Seymour would have wanted his creation to be used for, but as we've seen, inventions and innovations frequently take on lives of their own.

WHO MADE THE...
BARCODE?

...ne sky.

WHO MADE THE
NEON SIGN?
(see page 251)

...ade That?
...r curious minds.

Unravel how so many miraculous tools and technologies ended up right at your fingertips.

$19.95 US / $26.95 CAN

5 1995

ISBN 978-1-64643-215-8

9 781646 432158

You see them on everything, and by now, we just take them for granted. Barcodes have made buying and selling infinitely easier over the past fifty years, to where it's impossible to imagine commerce existing without them. But they've only been used in earnest since 1974, with the idea for them dating back almost thirty years prior.

In 1948, a food store chain in Philadelphia approached the Drexel Institute of Technology and asked if they could create a device or a system that could automatically read prices at checkout to speed up the process.

Bernard Silver and Norman Joseph Woodland stepped up to the plate. An early design used ink that was sensitive to ultraviolet light, but this didn't work so well. But the duo was able to acquire a patent in 1952 for a machine that was described as "article classification ... through the medium of identifying patterns." The problem was that technology was just not where the pair needed it to be at the time, and it took a while to catch up. Work on the design continued into the 1960s. In 1966, Silver and Woodland produced the first working barcode, but they realized that for the industry to adopt it, it would have to be subject to a set of universal standards, so that items could be used and read in any store that sold them. There couldn't be multiple versions or designs.

Woodland came up with the idea of a series of lines, the classic barcode, while lounging on the beach in Miami. He simply drew lines in the sand with his fingers and wondered if this could be the way to encode information about a specific product. Other designs included a "bull's-eye" design that Silver had favored and was put into production by Radio Corporation of America. Their design worked well during a test run at a Kroger grocery store in Cincinnati in 1972, but ultimately, a bar design submitted by IBM (based on Woodland's concept) won out, and was put into use in 1974. The first products to have the barcode were packages of Wrigley's gum.

The idea was slow to be adopted, but once more retailers saw the advantage of being able to track inventory, inhibit shoplifting, and speed customers through stores, resistance dwindled, and the barcode became ubiquitous.

WHO MADE THE...
CREDIT CARD?

Where would we be without them? The idea of giving people an option other than having cash on hand goes back to the late nineteenth century. At that time, some hotels and fancier stores starting issuing "charge coins." These coins were usually made out of metal and had a specific account number imprinted on them, which was kept on record at the establishment in question. They were given to valued customers to be used for purchasing, checking in, etc. One problem, though: these coins didn't have the bearer's name on them, so if they were lost or stolen, it would be pretty easy to run up charges on someone else's account.

In 1928, these coins started to be replaced by something more secure, the Charga-Plate. This small, dog tag–like plate had the person's name, city, and state embossed on it, as well as a small card on the back for an individual's signature. When a purchase was made, the plate was put into a machine with a slip of paper and the embossed information became imprinted on the paper, just as it would be on the carbon paper later used for credit cards.

By the 1930s, airlines were offering air travel cards that could be used on multiple airlines, allowing customers to "buy now, pay later," just like a modern credit card. Cards from Diners Club and American Express soon followed.

Finally, in 1958, Bank of America introduced its BankAmericard, the first true credit card. The program was launched in Fresno, California, and was well received. The bank went on to license the card to other banks, so that it could be used over a much wider area, and by 1966, it was nationwide. In the mid-1970s, the licensees agreed to use a new name, Visa, which has been with us ever since.

Meanwhile, in 1966, a group of California banks partnered to create the Interbank Card Association (ITC). The ITC issued its own rival card, Master Charge, which, of course, would eventually become MasterCard.

For better or worse, these little pieces of plastic have proven irresistible to consumers around the world. Over the decades, the cards have acquired magnetic strips to store information, chips that promise enhanced security, and the ability to facilitate contactless payments. What hasn't changed is the premise of buying now and paying later, with the cards effectively serving as small loans with high interest rates. Although they can come in handy in an emergency, it's safe to say that the credit card has been more helpful to the bottom lines of banks than it has to the public.

WHO MADE THE...
ELECTRONIC
CALCULATOR?

To the chagrin of students everywhere, math is essential to human civilization. A recognition of this reality are supported by the existence of ingenious devices like the abacus, which may have been invented in Mesopotamia almost 5,000 years ago.

But what about handheld personal calculators, the kind that can effortlessly multiply huge numbers, and accomplish other important tasks, like spelling out puerile words when you turn them upside down? Like many modern devices, these have their origins in inventions from the eighteenth and nineteenth centuries, when mechanical devices were created to help with addition and subtraction, and sometimes, multiplication and division.

The first electronic calculators appeared well into the twentieth century. In 1954, IBM created the first transistor calculator, a monster of a machine that took up half a wall. It could perform calculations, though, and quickly. It was available for people to buy at the reasonable price of … just over $83,000. Casio came out with a smaller version in 1957, though it was still embedded in a desk. In 1961, the British Bell Punch Co. released a true desktop calculator called the ANITA MK-8. It was about the size of a cash register, and it rather resembled one, too.

These innovations showed that these potentially handy machines could continue shrinking as computing power was increasing. In just a few short years, in 1967, Texas Instruments brought out the "Cal Tech" model, the first true handheld device. This closely resembled the contemporary idea of a calculator, but its outputs were strips of paper, which made it pretty impractical to carry around.

In 1971, that problem was solved by Busicom in Japan, who created the Busicom LE-120A, known as the "HANDY." It was pocket size and had a display screen for the results. It also cost $395. After that, the race was on to build newer, better, faster, and smaller devices that were affordable for everyone. Calculators began to be programmable and capable of complex mathematical functions, saving scientists and engineers huge amounts of time. Business calculators appeared, with the capability of performing math specific to accounting and other fiscal needs. These days, calculators can display graphs on full-color digital displays at resolutions comparable to phones. They've come a long way since those furniture-sized machines that cost more than most houses!

WHO MADE THE...
POST-IT NOTE?

These handy, little, slightly sticky pieces of paper have been used for everything from simple bookmarks to crucial reminders about that thing you need to do today—or else. The idea behind them is amazingly simple: a square or thin strip of paper that has a little bit of adhesive on one side, but not so much that it leaves behind any residue. They can cling virtually anywhere, at least for a while, though trying to reuse them usually ends up being a bit frustrating, if not outright comical as they fail and flutter to the ground.

The story behind the gummy bit on the Post-it is amusing. In 1968, a 3M scientist named Spencer Silver was trying to create super-strong adhesives. But he accidentally did the exact opposite: he managed to make an adhesive that stuck well enough, but didn't remain. It seemed like a cool idea, but no one was really interested. As he said, he'd created a "solution without a problem," that is, until another 3M scientist, Art Fry, had just the problem Silver was looking for.

Fry sang in a church choir and would use little pieces of paper to mark which hymns were to be sung. But, naturally, they kept on falling out, and he couldn't desecrate the hymnal by securing them with tape. What he needed were small pieces of paper he could write on that would also stick to the pages without damaging them. You can see where this is going, can't you?

Fry had heard Silver talking about his discovery of a marginally sticky substance, and he got the idea of applying it to small pieces of paper. He reached out to Silver, and the two got to work. Using pale yellow scrap paper, they worked out how to apply the substance so that the paper could then be stuck almost anywhere. Here was something that was useful for far more than a bookmark. You could leave notes about anything to anyone, anywhere!

They started using them at 3M and got fellow employees to adopt them, too. Everyone loved the little notes, so the company decided to roll them out in a trial. Initially marketed as "Press 'n' Peel," they didn't do so well. A year later, 3M tried again, offering free samples of "Post-its" to consumers in just one city: Boise, Idaho. This time, the reception was much better. People loved them, and over 90 percent said they would buy them if made available in stores.

From there, it was just a matter of marketing. Post-its took off in the 1980s, with everyone from individuals to major corporations snapping them up. Now these little strips of paper are sold in over 100 countries, and new versions are always being tested and released. But the originals, the classic yellow squares of scrap paper, are still a sentimental favorite.

WHO MADE THE...
ICE PACK?

Food spoilage has been an issue as long as humanity has been around. As have bumps, bruises, and sprains. The solution to both? Something very cold, of course! But before refrigeration was widely available, ice was not easy to come by in much of the world.

The history of modern ice packs can be divided into three eras: the early nineteenth century, the 1890s to the 1950s, and the 1950s to present. The march to these frozen delights began in 1805, when Frederic Tudor of Boston conceived of sawing large blocks of ice from the wintry covering of lakes and ponds in the region and preserving them in cold places to be used in hotter weather. His business model was successful, and he was actually able to export ice to other areas, carried by ship. In addition to being used to cool foods and preserve them, some doctors noticed that ice was useful to numb pain, or even prepare a patient for amputation (ouch!).

Selling blocks of ice became the norm in the nineteenth century, but by the 1890s, a patent was granted for a machine that manufactured ice, meaning that it would not have be "harvested" anymore. By 1903, another invention came along to help with pain relief: the rubber hot water bottle, invented by Eduard Penkala. Though hot water is the opposite of ice, it was used for similar medical purposes, and people began to see the value of using both hot and cold for various ailments.

From this developed the rubber ice bag, which, like the hot water bottle, was waterproof and had a stopper to prevent liquid from leaking out. But with the advent of electric refrigeration, scientists began wondering if something more efficient than "mere" water could be frozen, a substance that would take a lot longer to melt. In 1959, Albert A. Robbins invented an ice pack that would stay frozen for hours, allowing the user to keep food cold on picnics, long drives, and so on.

Then in 1971, Jacob Spencer, a pharmacist and sales rep for Pfizer, devised a different kind of pack, one that could be molded, making it more efficient to use on different parts of the body. It made use of a gel that could be frozen, to which he added the blue dye that is still common in these kinds of ice packs. Spencer did this, he said, for no other reason than that it made the product look more appealing.

Synthetic ice packs became the standard to keep food and drinks cool outside the home, and for calming down the effects of a bad sprain, keeping generations of picnickers and athletes happy.

WHO MADE...

GPS?

The Global Positioning System (GPS) technology is now so general that we hardly even notice it. And some find that a little unnerving. We can use GPS to plot a trip to an unknown location, to order a ride share, to make accurate web searches, or even just find the correct time. It's incredibly useful, even though it does have some creepy implications. But where did it come from?

The story begins with the Cold War and the Space Race. When the Soviets launched their Sputnik satellite in 1957, the Americans wanted a way to keep an eye on it. There was a simple means to do so: the Doppler effect. This means that the frequency of the satellite's radio signals increased as it was approaching and decreased as it was moving away. You can observe this same effect when a car approaches. If the window is down and the driver is playing music, you will notice that, as the car moves away from you, the music seems to get lower in pitch (because the sound waves have farther to travel before reaching your ears). This effect is also how astronomers know the universe is expanding, but that's another whole topic!

In any case, using this principle, scientists were able tell exactly where the Soviet satellite was. But after thinking about it a bit, they took the idea one step further. If they could locate the position of the satellite, they could also find the location of the receiver, based on how far it was

from that satellite. And so the idea of satellite navigation was born. The technology advanced throughout the
1960s, with American satellites being used to calculate
exact times and track the locations of nuclear subma-
rines, among other tasks.

By the early 1970s, the Department of Defense, eager to
improve U.S. tracking systems, developed a set of satellites to be used for navigation. Known as the Navigation
System with Timing and Ranging (NAVSTAR) satellite,
the first one was deployed in 1978, with twenty-four satellites being fully operational by 1993 (there are up to
five backup satellites, in case any of the main ones go
down).

GPS was initially used by the military, but its technology
is now widely available in cell phones and personal GPS
devices. The U.S. Air Force still monitors the system, and
pledges to have twenty-four satellites in use at all times,
helping the directionally challenged remain on track no
matter what.

SUPER SOAKER?

Squirt guns have long been a pastime during the hot days of summer. They are a simple path to plenty of fun: take a humble piece of plastic, fill it with water, and pull the trigger to unleash a stream of water. The only problem was that they tended to go through water quickly—once you stopped to refill, you inevitably got blasted by your opponents.

Since no one is ever content to leave well enough alone, the search for a better squirt gun was on! In 1982, a NASA engineer named Lonnie Johnson discovered one while working on improving refrigeration aboard spacecraft. Johnson wanted a heat pump that used water, so he hooked up his prototype to the bathroom sink. As for what happened next, he said: "I turned around and I was shooting this thing across the bathroom into the tub, and the stream of water was so powerful that the curtains were swirling in the breeze it sent out ... I thought, 'This would make a great water gun.'"

Johnson put together a prototype, but quickly realized he didn't have the means to mass produce it. So he started seeking out existing toy manufacturers. It took several years for him to find a good partner; several companies were interested but, perhaps influenced by the public's growing displeasure with toy guns, couldn't go through with it. In 1989, Johnson happened to meet the head of the Larami toy company, who showed genuine interest. Johnson and the designers at the company worked to refine the design, and in 1990, it was launched as the Power Drencher. A year later, it was rebranded as the Super Soaker, a name that has been used ever since. It doesn't really resemble a real gun, but has a big water bottle on top that allows the user to shoot a whole lot of water before refilling.

The best-selling model is known as the Super Soaker 50, though there have been many other versions along the way. The SS50 is considered important enough that it was nominated for a place in the National Toy Hall of Fame. Not bad for a repurposed refrigeration pump!

WHO MADE THE...
CROCKPOT?

The beloved Crockpot was born out of a need created by a shift in society, and fueled by nostalgia. During World War II, many American women worked long hours away from their homes to help with the effort, and although many returned to the home after the war, some remained in the workforce. But, due to the rigid gender roles in place at the time, these working women were still expected get dinner on the table each night. The Crockpot, then, was a considerable help, as they could put everything in it in the morning and let it slowly cook throughout the day so that it would be ready when they and their husbands returned in the evening.

But the actual origins of the Crockpot go back even further, to the late nineteenth century. In Vilna, a Jewish neighborhood in the city of Vilnius, Lithuania (a neighborhood once known as the "Jerusalem of the North"), a tradition arose where the residents would make a meat-and-bean stew on Fridays before the start of the Sabbath at nightfall. Families would prepare these stews in crocks and take them to local bakeries where the ovens, although they had been shut off for the day, were still hot and cooled slowly overnight. The crockery was placed in these ovens and allowed to cook slowly overnight, meaning that the stew inside, *cholent*, would be ready by Saturday, and a meal could be enjoyed without anyone having to work to prepare it. It was an ingenious solution to a longstanding dilemma for the Jewish faithful.

Several decades later, one Irving Nachumsohn (later Irving Naxon) created the first slow-cooker, basing his design on his mother's descriptions of Vilna's tradition. Working as an engineer for Western Electric, Naxon loved tinkering on the side and came up with several inventions during his tenure. But the one he will most be remembered for is the Crockpot.

The first slow-cooker was called the "Naxon Beanery," and it was shorter and squatter than later models. While it sold well enough, the handy device really started to take off when Naxon sold his company and the patent to the Rival Company in 1970. Rival's designers developed new models, rebranded the cooker as the "Crockpot," and, with the help of inventor Robert Glen Martin, focused on making it able to cook whole meals, not just stews.

Before long, other companies were producing their own versions, and the popularity of the device exploded. It has the advantage of helping people who don't like or don't want to cook produce a delicious meal with minimal effort. It's also very beginner-friendly, a major part of the Crockpot's enduring appeal.

WHO MADE...
SILLY PUTTY?

Silly Putty is a delightfully weird substance that has been entertaining children for generations. A lump of sticky, clay-like material, it can also be imprinted with images, such as words or cartoons from a newspaper. And, appropriately enough, it was discovered by accident, though by who remains a bit of a mystery.

The invention (or rather, discovery) of Silly Putty could be attributed to Earl Warrick and Rob Roy McGregor at Dow Corning, or Harvey Chin, or James Wright at General Electric in New Haven, Connecticut. Warrick insisted that he and McGregor discovered the substance and received their patent first, but the Crayola company insists that it was Wright. In any case, both parties seemed to have discovered this unusual material in 1943—just another one of those zeitgeist things, this time just for something weird and, well, silly.

Both camps found that mixing boric acid with silicone oil produced an odd substance. It was more elastic than rubber. It could be formed into a ball and bounced against a hard surface. It didn't melt, and it didn't spoil. Initially, there was some hope that this material could be a synthetic rubber, a very useful substance to have during World War II, when rubber shortages were common. But it proved not to be a very good substitute.

Wright contacted other scientists to see if they could come up with good uses for it, but this "Nutty Putty" wasn't a big hit.

At least not until 1949. A toy store owner, Ruth Fallgatter, discovered the putty, and bought some as a novelty. It sold, but she didn't see a grand future for it. Her marketing consultant Peter C. L. Hodgson, however, thought differently. Already in considerable debt, he borrowed some money and packed the substance into plastic eggs, calling it "Silly Putty." Not long after, the *New Yorker* did a small feature on it, and sales went through the roof, with 250,000 eggs selling in just three days. Hodgson's financial issues were suddenly past, but his business almost ended when in 1951, the U.S. government imposed a ration on silicone, since it was needed for the Korean War effort.

Fortunately, this ration only lasted for a year, and production of the Putty resumed. At first, it seemed like adults were more interested in it than kids, but that soon changed, especially after a commercial for it aired during the *Howdy Doody Show* in 1957.

Silly Putty became a worldwide phenomenon soon after—even the Soviet Union couldn't keep it out. In

1968, the astronauts on Apollo 8 took some into space. But they weren't playing with it—instead, it was used to hold tools in place in zero gravity. Silly Putty has been with us ever since, and has also found uses such as in physiotherapy, where it can be useful for hand injuries and improving grip strength. Its current owner, Crayola, still sells millions of units a year around the world, and it was inducted into the National Toy Hall of Fame in 2001.

WHO MADE THE...
SLINKY?

The Slinky is a wonderful toy that takes on a life of its own as it meanders down the stairs. Its construction is ingeniously simple, nothing more than a tightly coiled spring. But how the Slinky came about is far more interesting, and like a lot of successful inventions, it happened by accident.

In 1943, an American naval engineer named Richard James was developing ideas for springs that would keep important and sensitive equipment safe from the turbulent sea, essentially serving as a shock absorber. While stationed at a shipyard in Philadelphia, James created a tightly coiled spring. One day, he accidentally knocked one of these off the shelf and was astonished to see that it landed and proceeded to "walk" onto to some books stacked near the shelf. The spring then continued down to the floor, where it coiled up and stood upright, looking exactly as it had before he knocked it off.

James went home and told his wife, Betty, and proceeded to refine the spring into a fun toy. Betty came up with the name "Slinky" after looking through the dictionary and happening upon the word, which means "slick and graceful." Richard tinkered with the design for a while, and then, when he felt he had perfected it, he and Betty formed James Industries. Taking out a $500 loan, they had 400 Slinkys manufactured. As with so many inventions, they initially had trouble convincing local toy shops that the product would be a hit.

But they were given a chance to demonstrate the Slinky at a display at Gimbels department store in Philadelphia in November 1945, just as the Christmas shopping season was commencing. Customers were wowed by the novel toy, and all 400 sold out in less than two hours! Richard and Betty knew they were onto something. Over the next few years, Richard continued to fine-tune the design and develop related products. He licensed the patent to other companies, and sales took off. It's estimated that in the first few years of full-on production, 100 million Slinkys were sold at $1 each. Yep, $100 million in Slinky sales, which according to calculations, works out to about $6 *billion* in today's money!

Slinkys continued to sell well, and after Richard and Betty divorced in 1960, she took over the company (he decided to dedicate himself to missionary work in South America). Betty would stay on as president of James Industries until 1998, when, at the age of 80, she stepped down. But her departure couldn't stop the momentum of the Slinky—it continues to delight children and adults all over the world.

WHO MADE THE...
THUMB DRIVE?

The thumb drive is an amazingly handy little storage device. If one thinks about how much the technology has progressed even in the last ten years, it's rather mind-boggling. These tiny little sticks can now hold vast amounts of data, far more than was even thinkable a decade ago, and at reasonable prices. They are in use everywhere, helping individuals and gigantic corporations. Just make sure you don't misplace yours ...

But as for who invented it? Well, that's a bit of a quagmire. Inventors Amir Ban, Dov Moran, and Oron Ogdan of the Israeli company M-Systems filed a patent application in April 1999 for their "flash drive." But in that same year, an IBM engineer named Shimon Shmueli submitted what is known as an invention disclosure, where he asserted that it was he who had invented the USB flash drive. Not to be outdone, Trek 2000 International in Singapore was the first to sell these drives, and it insists that its people invented it. Oh, and Malaysian inventor Pua Khein-Seng claims that it was his invention. So honestly, who knows?

As you might imagine, there have been several rights and trademark lawsuits over the years, with multiple parties trying to stake their claims as inventors. But the issue doesn't seem to have been settled, even more than twenty years later. So, pick whichever inventor's version you like best!

As long as the things work as advertised, most consumers probably don't even care. And work they do. The capacity for storage in these little devices is astonishing. The first thumb drives were only able to hold a whopping 8 megabytes, though when you remember that PCs in the 1990s routinely only had hard drives in the 300- to 400-megabyte range, this doesn't seem too paltry.

With the inevitable advances in technology over the past two decades, it's now possible for these little drives to hold an incredible 2 terabytes of data. That's 2,000 gigabytes, or 2 *million* megabytes! The storage capacity has come so far that one wonders just how much further the technology can be pushed. If history offers any clue, the answer might be unfathomable to us where we stand today.

WHO MADE THE...
CARABINER?

The word for these useful little clips derives from the German *karabinerhaken*, or "spring hook." They have their origins with the carabinier, a class of soldier in the French army during the mid-seventeenth century. They were mounted cavalry that carried shorter versions of the standard rifles of the time. This shorter rifle was known as a *carbine*, or carabine in French. These weapons allowed the soldiers options during an attack—they could either fire from horseback, or after dismounting. But to use their rifles effectively, they needed to be able to hold on to them (and not drop them while riding), so they were attached the soldiers' belts with a strap and a pair of clips. These clips were the forerunners of the versatile hooks that mountain climbers and others now rely on.

Fast-forward to the early twentieth and a mountaineer named Otto Herzog, whose nickname was "Rambo," long before Sylvester Stallone took ownership of that moniker (the name derives from the German word, *Ramponieren*, which means "to batter" or "to bash"). Apparently, Herzog earned this name through his sheer determination and willpower. He introduced the idea of a modern carabiner as an aid in mountain climbing, a very strong and reliable hook that could attach to one's belt. If there's any quality you want when halfway up a mountain, it's reliability! "Used once" is not a good selling point for used mountain climbing equipment.

Herzog seems to have come up with the idea for a new, improved carabiner when he saw some firefighters in 1910 wearing similar clips on their belts. If firefighters trusted them, why shouldn't mountain climbers? Only Herzog's early efforts didn't work as well as he liked; the clip didn't stay closed as well as he wanted, a major problem when one's a few hundred feet up in the air.

Later, in 1938, two climbers—Paul Allain and Raffi Bedayn—both came up with the idea of carabiners made of aluminum, a material that allowed the clips to be both lighter and stronger. Who had the idea first? It's another one of those zeitgeist things, in that both seemed to have conceived of the concept independently. In the long run, Bedayn's model won out, becoming a standard piece of climbing equipment over the decades. There have been innovations in style, color, and structure since then (some even double as bottle openers or screwdrivers, for example), but the basic clip is pretty much the same, and has made scaling cliffs and mountainsides safer and more enjoyable for countless folks ever since.

WHO MADE THE...
MICROPHONE?

Where would we be without the microphone? Well, a lot less cultured, since everything we hear, whether live or recorded, depends upon them. All our favorite recorded music, movies, TV shows, and many other cherished forms of media would not exist without a means of accurately capturing and conveying sound to recording equipment. Come to think of it, recording equipment wouldn't exist, either.

A microphone can also amplify the human voice, a desire that likely dates to the popular dramas of ancient Greece, when certain masks worn by the actors featured mouthpieces in the shape of horns to allow their voices carry better. But, obviously, the idea of using an electronic device to capture sound so that it could be amplified or recorded would not come until much later.

By the 1860s and 1870s, several individuals were focused on creating a machine that could transmit sound. Obviously, this went hand in hand with the telephone (see that entry for more). Indeed, Alexander Graham Bell patented a prototype microphone as early as 1876. A major improvement came with the invention of the carbon microphone in 1877. So who came up with it? Well, here we go again! At least three inventors claimed to have created the first such device: the Englishman David Edward Hughes, and Emile Berliner and Thomas Edison in the United States. Although Edison jumped on procuring the patent in 1877, Hughes almost certainly came up with the idea first, having demonstrated his own prototypes a few years earlier. Edison's gambit is in line with his documented behavior during this time, as the "Wizard of Menlo Park" seems to have been fairly unscrupulous in regard to claiming inventions as his own and then trying to back these claims up officially.

In any case, the carbon microphone worked far better than Bell's offering; without the giant leap forward it made, one has to wonder where society would be. The next step in microphone evolution came in 1916, with the invention of the condenser or capacitor microphone by E. C. Wente. These microphones allowed for sound to be converted into electrical signals. As technology progressed, and the desire for better and clearer recordings and broadcasts grew, innovations in the microphone sphere continued apace. Ribbon microphones, candlestick microphones, and many others appeared in the 1930s and '40s. Radio broadcasts and vinyl recordings became all the rage during this time, and microphones had to keep up. Several new companies sprung up, each trying to tweak existing designs to meet the considerable demand and outdo their competitors. Innovation continues to this day, making mics that feature digital recording technology, fiber optics, lasers, and other like-magic technologies available to anyone at the click of a few buttons.

WHO MADE THE...
DOG LEASH?

Dogs have been companions to humans for a very long time— the domestication of dogs seems to have sprung up in multiple places across Europe and Asia between 29,000 and 14,000 years ago. It is thought that this development began with packs of wolves following hunter-gatherer groups, taking the leftovers from humanity's hunts and gradually building up affection and trust between animals and humans.

Over time, the two groups may have even begun to hunt together. At some point, it likely became necessary for people to restrain the wolves, and so rudimentary forms of the leash might well have been formulated way back when. But the first evidence of a leash dates from about 8,000 years ago, in rock art from Shuwaymis, a town in what is now northeastern Saudi Arabia. The panel shows a pack of dogs, two of which are connected to a man (presumably a hunter) by a line.

Later, in Mesopotamia, artwork clearly shows dogs with cords around their necks, by which the owner was able to control them. This simple design seems to have endured for centuries, with variations on it showing up in numerous civilizations, including Egypt, where dog collars have been found in tombs, some even with the dogs' names inscribed. By the seventh century

BCE, dogs in Assyria owned by upper-class individuals often had ornate collars and leashes to reflect their owners' wealth.

From this time forward, dog leashes could be found, in one form or another, virtually everywhere in the world. There is no one certain inventor of the dog leash, just a succession of people who came up with variations on an old idea. Whether a simple rope or a leather strap studded with gold and jewels, the leash may not be quite as old as the emotional bond between dogs and humans, but it likely followed shortly after.

WHO MADE THE...

FIDGET SPINNER?

The fidget spinner is one of the most recent inventions covered in this book. At the time of writing, it wasn't even five years old. But the idea behind it goes back to 1993, when a chemical engineer named Catherine Hettinger came up with an idea for a spinning, top-like toy that could be held in the palm of one's hand. Hettinger's concept looked nothing like the modern fidget spinner, and instead resembled a small UFO that could be balanced upon a finger. She obtained a patent for the toy and tried to interest companies in her design, but no one bit. Faced with no interest, Hettinger let the patent lapse in 2005. That should have been her exit from the story, but some news outlets later, erroneously, tried to connect her with the current iteration, a connection that Hettinger herself dismissed.

But yearning for a spinning, handheld toy endured. Scott McCoskery came up with his own design for a spinning device over the course of 2013 and 2014. He said it was to keep himself busy and prevent fidgeting at meetings. He initially called it a "Torqbar," and there was interest via online sales. Of course, as soon as some buzz started building, other companies began ripping McCoskery's concept off. He modified the design, and by 2017, it was selling as the Fidget Spinner we now know and love. And sell it did! The little devices were a smash hit, as fidgety people all over the world wanted to give themselves something to do when tedium descended.

Marketed as a stress-relieving toy, they are just as popular (if not more) with adults as they are with kids. Other companies, such as Disney, saw the benefit of having their own brands featured on the spinners, and set about making their own. There is no patent on the Spinner, so companies are free to make their own versions and innovate. Scott's not complaining, though. His initial offering was a massive success and is still sought after by legions of nervous types.

WHO MADE...
LEGOS?

Legos are one of the most recognizable products in the world.
Originally created as a simple but fun system of building blocks, modern Lego sets can be extremely complex, producing structures that are nothing less than works of art. Lego is also one of the most valuable brands in the world, beloved by children and adults everywhere. But how did this mighty empire begin? With one man in a small town.

Ole Kirk Christiansen was a carpenter from Billund, Denmark, who began making wooden toys in 1932. By 1935, he had established his company, Lego. The name comes from the Danish words *leg godt*, which translate to "play well." At the time, Christiansen didn't realize that the word "Lego" also means "I put together" in Latin. A happy accident, to say the least!

Throughout the 1930s, he and his son, Godtfred, built wooden toys. Though their work was then interrupted by World War II, the company was producing what it called "Automatic Binding Bricks," based off a British design, by the end of the 1940s. These were the forerunners of the Legos we know and love today. During this time, the company was moving more and more toward plastic, despite public opinion not being in favor of plastics replacing more traditional materials. As Ole was known for insisting on very high quality for all of his products, this seems to have won over some anti-plastic crusaders.

In 1954, Godtfred began to oversee the day-to-day operations of the company, and it was under his direction that the move toward the modern Lego began. Following conversations with his father, Godtfred came to see the value in an improved system of interlocking bricks, and set about developing it. The idea was to provide structure so that the more bricks you had, the more elaborate structures you could build. It took several years, but a patent for the modern Lego brick design was filed in January 1958, and shortly after, the rollout began.

After a fire destroyed the company's wooden toy manufacturing plant in 1960, Lego focused exclusively on its plastics. This proved to be a wise choice, as the bricks became very popular. Their versatility was a big part of the reason. Legos were endlessly usable, expandable, could be played with anytime, were for both boys and girls, stimulated the imagination, were durable, and were of high quality.

Lego continued to develop new and more complex designs. It introduced its first figures in 1978, and these have proven immensely popular, so much that they managed to power the critical and box office success of the Lego movies. The bricks have been used to build detailed scale models and even 1:1 full-scale models of objects such as the X-wing fighter from the Star Wars franchise. These developments are all a far cry from what a toy maker in a Danish town first envisioned, but show the potential present in a great idea!

WHO MADE THE...
NERF GUN?

Nerf balls initially became popular in the 1970s. Super soft and made of foam, they could be played with inside and present (hopefully) little to no threat to things like vases and framed pictures on the wall. Where basketballs and baseballs might be impractical and dangerous in the house, the Nerf ball was the perfect solution for kids stuck inside on a rainy day.

The idea for these less-bruising balls originated with Reyn Guyer, a game inventor (he also created Twister) who approached Parker Brothers in 1968 with an idea for games that could be played safely indoors with a foam ball. Parker Brothers wasn't interested in the games, but they loved the idea of the ball itself, and began marketing the Nerf ball as safe for indoor use. It was a smash hit (so to speak), and by 1972, Parker Brothers returned to Guyer, asking him for some of the games he'd originally proposed. One of the first produced was the Nerf football, which worked just as well outdoors as it did inside, being easier for young hands to grasp.

The company dipped its toe into other games, such as Nerf versions of Ping-Pong and pool, but none had the popularity of the football—until the Nerf gun arrived. In 1989, Nerf introduced the Blast A Ball, a small, cannon-like device that shot soft, golf ball–sized spheres. Not content with this, the company decided that what it really needed was to create a line of toy weapons that resembled objects from a futuristic sci-fi story, but shot soft, foam darts. Some shot darts with Velcro tips that could stick to a Nerf vest worn by one's opponent, while others had suction cups allowing them to stick to smooth surfaces.

The company has created a variety of variations on this theme, including models such as N-Strike Elite, Vortex, Rebelle, Zombie Strike (of course!), Ultra, and the post-apocalyptic Doomlands 2169 (wielders of the gun must fight off strange monsters that haunt the world after an asteroid strike). Nerf guns have breathed plenty of life into a classic toy, but they're a far cry from Guyer's pacific ideas for indoor fun!

WHO MADE THE...
ETCH A SKETCH?

These fun drawing pads have delighted children and adults for years. Although they can be great for making light-hearted doodles on, some artists have used them to create very impressive artworks, even if they are only temporary. The cool thing about the Etch A Sketch is its self-cleaning feature—with a simple shake, your work is undone, and you have a "blank canvas" once more.

The Etch A Sketch was invented by a French electrical technician, André Cassagnes, to explore the clinging properties of electrostatic charges. Using a screen coated with aluminum powder, he was able to create a novel kind of drawing board that would leaving marks behind when scratched by a stylus. He called it *L'Ecran Magique*, or, "the magic screen," and presented it at the International Toy Fair in Nuremberg, Germany, in 1959. The Ohio Art Company saw Cassagnes's product and decided to bring it to market in time for the 1960 holiday season. Renaming it "Etch A Sketch," they shifted into high gear manufacturing and marketing it.

Their faith was well founded. The little drawing boards were a massive hit, and continued to be a must-have toy in the years that followed. However, there was a design problem: the screen was made of glass, and many safety advocates worried that this posed a threat to children, because they are not always the most careful users of things. In 1970, the Consumers Union asked the Department of Health, Education and Welfare to take a look at these concerns, citing a newly passed law concerning toy safety and child protection.

After looking closer at the Etch A Sketch, the department ordered the company to redesign it with a plastic screen. Happily, this proved relatively easy, and before long, the little tablets were out in the world again, this time in a safer package. Other versions have appeared since then, including a color version, and one allowing for video game–like play and animation, but the original endures, providing a bit of artistic fun for all ages.

WHO MADE THE...
HULA-HOOP?

The hoop is a very old instrument that has been used for numerous purposes over the centuries, one of which was amusement—there are records of young folks in medieval Europe having fun with hoop-based toys. So how did the Hula-Hoop explode on the scene in the twentieth century, if there's very little novelty to it?

The phenomenon started in 1957, when a woman named Joan Anderson returned from Australia with a bamboo exercise hoop she'd seen children using in a gym class. Her husband showed it to Arthur Melin and Richard Knerr, founders of the toy company Wham-O (who had already produced the wildly popular Frisbee). Melin was intrigued and made a handshake agreement to share profits with Joan and her husband. Melin then, of course, went back on his word and cut them out. The moral of this story: always get it in writing!

After this shady beginning, Wham-O began manufacturing a plastic version of the hoop, which it christened the "Hula-Hoop." The name "Hula" was chosen because the Polynesian craze that swept America in the middle of the twentieth century, and a misconception about how the waist functioned in traditional Hawaiian dance. So yes, these guys were pretty problematic on several levels, but, as it was the pre-woke culture of the '50s, they got away with it.

In any case, the Hula-Hoop was a smash success, with something like 25 million selling in the first four months after its release in July 1958. The sales remained strong over the next couple of years, but alas, like any craze, the demand eventually died down, and the Hula-Hoop took a bit of a back seat in the culture.

However, hooping was taken up by circus performers and other acrobats, especially in China and Russia, where skills with the toy reached incredible new heights. It wasn't just a matter of twirling the waist; now hoopers were doing stunning tricks and feats with them, often using multiple hoops at once. And new designs were emerging, ones better suited to this growing athletic mastery. Hooping has become an essential part of circus arts communities, Burning Man, fire performances, and the like, showing no sign of slowing down.

WHO MADE...
LINCOLN LOGS?

Interestingly, these rustic cousins to Legos and Tinkertoys were not originally intended to serve as a children's plaything. Instead, they were inspired by something deadly serious. John Lloyd Wright, son of the famed architect Frank Lloyd Wright, was working with his father in Tokyo on the Imperial Hotel, a project that began in 1916. Knowing that the region was prone to earthquakes, Frank came up with an ingenious solution: a series of interlocking timber beams that would create a more stable structure. The building would sway during an earthquake, but not collapse. The design proved to be very effective. In 1923, the Kanto earthquake severely damaged Tokyo, but the hotel, using the great architect's design, remained standing.

Though John and his famous father had a falling-out during the Imperial project, he loved his father's idea, and came up with the idea of utilizing the interlocking design as a toy. Creating the Red Square Toy Company, he began marketing it in 1918. The idea was that not only were these sets fun to build with, but they could withstand a typical child's roughhousing and not fall over, unlike building blocks. Receiving a patent for his concept in 1920, John changed the company name to J. L. Wright Manufacturing. He then capitalized on the nostalgia for the log cabin by deciding on the name "Lincoln Logs," evoking the well-known story that Abraham Lincoln was born in one and. Early Lincoln Log sets included instructions on how to build both Lincoln's mythical cabin and the cabin from the famous novel *Uncle Tom's Cabin*. The box featured the slogan "Interesting playthings typifying the spirit of America." In post–World War I America, it was an irresistible toy for many, and marketed to both boys and girls.

Lincoln Logs continued to be popular over the next several decades, and didn't suffer the restrictions on manufacturing that some other toys did during World War II. While the toy's popularity peaked in the 1950s and 1960s, these logs are still being made today and continue to have a number of dedicated fans.

WHO MADE...
PLAY-DOH?

This squishable, colorful substance has been a childhood favorite since the 1950s, but it started out as anything but child's play. No, Play Doh's origins are decidedly more unusual: it was originally intended to be used as a wallpaper cleaner. Can you think of anything less thrilling?

The story goes back to 1933. Cleo McVicker worked for a soap company called Kutol Products in Cincinnati, Ohio. Unfortunately, sales were not going well and Kutol was on the verge of going out of business. But McVicker had an idea. He contacted a local grocery chain, Kroger, and negotiated a deal to make a wallpaper cleaner that Kroger would sell in its stores. There was only one problem: Kutol didn't make wallpaper cleaner.

They put their heads together and came up with a simple, clay-like substance, made primarily from water, flour, and salt (hence the "dough" connotation in the toy's eventual name) that was nontoxic and good at removing coal-smoke stains from wallpaper. It was enough to save Kutol and keep them in business for another two decades. But in the aftermath of World War II, a large number of houses began to convert their heating to gas and oil, rather than relying on coal. So Kutol was in danger of going under yet again. Once again, they adapted.

Cleo's son Joseph was now running the business. His sister-in-law, Kay, was a schoolteacher who saw the value in using the substance as a safe way to teach crafts in schools. Joseph became intrigued once he saw what was happening in classrooms that were already using it. He created a new subsidiary of Kutol, called the Rainbow Crafts Company, and set about manufacturing the stuff for an entirely new purpose.

The name *Play-Doh* came from Kay and her husband, Noah. Joseph and Noah had wanted to call it Rainbow Modeling Compound, so let's be glad Kay won out! Once available, Play-Doh was gaining popularity, but Joseph knew that it needed a bigger audience to truly succeed. He reached out to Bob Keeshan, who loved the stuff and wanted to feature it on his television show. The two agreed that it would be used on the show at least once a week. Oh, Keeshan's better-known name? Captain Kangaroo. Yep, Joseph had scored free publicity with one of the most popular kids' television shows of the 1950s.

As you can imagine, sales went through the roof. In the 1960s, the company developed the Fun Factory concept, which allowed dough to be shaped with simple tools. Play-Doh has never looked back since. It's estimated that since its debut in the mid-1950s, well over 2 *billion* cans have been sold!

WHO MADE...
ROLLER SKATES?

Shoes with wheels seem like either a really cool idea or a really bad one. For those able to master them, roller skates are very cool indeed. The idea is fairly simple, but for whatever reason, no one seems to have thought it up until the mid-eighteenth century. In 1743, some London-based actors attached wheels to their shoes to mimic the action of ice skating for the production they were in. But this innovation wasn't the start of a new craze. The first attempt at genuine roller skates came a few years later, when Belgian inventor John Joseph Merlin crashed the party ... literally.

In 1760, Merlin attached wheels to his shoes and tried to impress people at a masquerade party by skating through the room while playing a violin. There was only one problem: he had no way to stop his momentum. Also, his balance wasn't all that good, as he was about to painfully find out. Unable to slow down as he skated past the undoubtedly astonished partygoers, Merlin crashed into a mirror, shattering it, breaking his violin, and injuring himself.

That debacle should have been the end of roller skates, and for a while, it pretty much was. A three-wheeled version appeared in Paris in 1819, and other stage productions used actors in wheel-enhanced shoes to simulate ice skating, but it wasn't until 1863 that inventor James

Plimpton of Massachusetts came up with the idea for the "rocking" skate, an innovation that allowed the shoe to tip forward and act as a brake. Plimpton also introduced the four-wheel model that is still in use.

The roller-skating craze caught on, and soon rinks were popping up throughout the Northeast. "Rinkomaina" became a thing, especially among the young. Indeed, roller skating was seen as an acceptable way for male and female teenagers to interact without requiring chaperoning (the late nineteenth century being what it was). Some physicians also began to recommend skating as a good and safe form of exercise. In 1885, *Scientific American*, in wonderfully mannered prose, said: "In proportion to the immense number of persons who have been engaged in propulsive divagations upon polished floors during the past winter, the pathological outcome has been small." Not exactly ringing endorsement!

Skating didn't wane in popularity as the new century dawned, and in time, new forms of skates and whole new sports popped up, from roller derby to rollerblading. But we can thank James Plimpton for setting the whole thing in motion, literally.

WHO MADE THE...
YO-YO?

Yo-yos have been around for quite a long time. They may have first been known in China, but the best concrete evidence we have for them in antiquity comes from ancient Greece. A vase from ca. 440 BCE clearly shows a boy playing with one. It was thought that ceramic (and possibly nonfunctioning) versions of yo-yos were offered up to the gods as a sacrifice when boys came of age, a sign that they were putting childish things behind them. As a toy, the yo-yo was largely forgotten in the West after the classical Greek age passed. But it has long been popular in the islands now known as the Philippines, lending credibility to the theory that the toy originated in Asia.

By the eighteenth century, the yo-yo seems to have worked its way back into favor with the upper classes of Europe. The yo-yo was especially popular in France, where, in some circles, it was called an *emigrette*. One theory for this name is that the toy was popular with both adults and children of the aristocracy, many of whom were forced to flee or "emigrate" out of France during the French Revolution. In any case, the yo-yo survived those turbulent times with its proverbial head intact and went on to become popular in many other nations, including Germany and England.

Yo-yos were not just seen as toys, but also as stress relievers, and there are almost as many images from these times of adults using them as children. Various patents for improvement to the designs were taken out in the later nineteenth century. In 1911, *Scientific American* ran an article about Filipino toys that included the word "yo-yo" for the first time. The word probably comes from yóyo, a word in the Ilocano language that means something like "return."

In 1928, a Filipino immigrant to the United States named Pedro Flores established the Yo-yo Manufacturing Company in Santa Barbara, California, and he is credited with playing a major role in popularizing the toy among the American public. Indeed, in less than a year, his factories were making as many as 300,000 yo-yos per day!

Around that time, a businessman named Donald Duncan recognized the yo-yo's potential to become even bigger. He not only bought Flores's company, but also the name. Duncan was a very good businessman, and the product only grew under his direction. He trademarked the name "yo-yo," thus preventing competitors from using it. And while his company ran into trouble in the 1960s and was purchased, the "Duncan yo-yo" is still a recognized brand around the world.

WHO MADE THE...
JUMP ROPE?

The jump rope is a beloved pastime for kids, provides a great workout for athletes, and is an important part of some dance styles, such as hip hop. For some, it's easy and fun, for others, it's something they'll never quite get right. It can be used on your own or in a group, with one rope, two, or even more!

But who invented it? It may be that jumping rope started in several places simultaneously. As we've seen, people can be completely separated from one another and still come up with the same idea.

There is evidence that in ancient Egypt athletes would improve their coordination by jumping over vines that were stretched out or shaken up and down. The Aboriginal peoples of Australia might have done the same. At some point during these exercises, people might have discovered that this form of physical training was also a whole lot of fun.

Other accounts claim that the true origins of jumping rope lie in Asia. The Traditional Chinese Game League says that in ancient China a game was played at Chinese New Year called "jumping 100 threads," because, when twirled quickly, a single rope can look like many ropes as it circles through the air. Japanese soldiers might have been using a form of jump rope as part of their training over 2,000 years ago.

There are other theories. The National Double Dutch League says that its form of jumping rope, Double Dutch, might have begun with the Phoenicians, a culture that thrived over 4,000 years ago in what is now modern Lebanon. This style of jumping rope features two people standing at each end of two ropes, turning them in opposite directions. The story goes that Phoenician rope makers would back away from each other while twisting stands of hemp to make rope. Assistants would help them out by giving them more material and going back and forth between them. They had to learn to jump over the rope as it became longer. The name "Double Dutch" might have come from adults seeing Dutch children in seventeenth-century New Amsterdam (a town that later became New York) playing a version of the game.

The problem with stories like this is that we don't really know which, if any, are true. They might all be, and it's entirely possible (even likely) that a game or a sport as simple as jumping rope was "invented" and "discovered" many times around the world over the centuries. We do know that in Europe, beginning around the year 1600, there are surviving drawings of children playing jump-roping games. That doesn't mean these games were invented then, of course; children could have been playing them for centuries. So wherever the jump rope came from (quite a lot of places, it seems!), it was pretty well established by the beginning of the seventeenth century and would never cease to be popular from then on.

WHO MADE...

MR. POTATO HEAD?

The weird and wonderful Mr. Potato Head and his friends have been entertaining folks for over a half a century. It's an easy idea to comprehend, but one that seems a bit strange upon closer examination. Why would anyone want to put facial features, arms and legs, and other accessories on a potato? And why a potato?

The story of this spud started with an inventor named George Lerner. In the 1940s, Lerner came up with the idea of creating strange little figures and dolls for his younger sisters to play with. He hoped that there might be a market for them, but during the World War II era, using food for anything other than eating was seen as wasteful, so Lerner's idea didn't take off.

Still, Lerner stuck with the idea of a toy that children could personalize by using a variety of accessories: eyes, mouths, arms and legs, hats, even a pipe. He was able to convince a food company to include these accessories in breakfast cereals as prizes, a "collect them all" kind of promotion. Eventually, Lerner attracted the attention of a toy company called Hassenfeld Brothers (later the more well-known Hasbro), who saw his idea as unique, to say the least!

Hasbro bought the rights and began marketing the accessories in bundles, where you would have everything to make funny figures out of whatever veggies you had at hand. And that was the thing; the early versions of the toy didn't come with a plastic potato. Instead, there was a Styrofoam head kids could use, or parents could provide a real vegetable: a potato, a cucumber, an eggplant, whatever.

Mr. Potato Head has the distinction of being the subject of the very first television ad aimed at children, rather than adults. After it premiered in April 1952, the toy took off and sold more than a million kits over the next year. Wanting to capitalize on this, Hasbro added more toys to the line: Mrs. Potato Head, Brother Spud, and Sister Yam. Then came the accessories: a car, a boat trailer, a kitchen, and even pets for the family.

Changes in U.S. safety laws forced Hasbro to make the parts less sharp, which meant that they no longer could be affixed as easily to real veggies. To reckon with this reality, the company came up with the plastic head we know and love in 1964, and added a few others, such as Oscar the Orange, Pete the Pepper, Cooky the Cucumber, and Katie the Carrot. The weird little idea became a craze. The toys were made even more popular in recent years by appearing in the Toy Story movies, igniting the obsession for a whole new generation.

WHO MADE THE...
JIGSAW
PUZZLE?

Jigsaw puzzles can be easy and fun or frustrating and difficult, depending on how many pieces you're dealing with and what the final image is. They're a great way to pass the time, offering a genuine challenge and a wonderful sense of satisfaction when they're completed.

Although the idea is so simple it seems like they could have been around for a long time, the first known jigsaw puzzle was invented in London in 1762 by an engraver and cartographer named John Spilsbury, who mounted a map on wood and used a saw to cut along the national boundaries. The point was to use the puzzle as a way of teaching children geography. Spilsbury's idea even caught on with the royal family; King George III had geography puzzles made for his children.

Known as "dissected maps" or "dissected puzzles," they were used mainly as teaching aids for several decades. But eventually, they became popular as a leisure activity with adults and children. By the 1880s, these puzzles were created with a tool called a fretsaw. For reasons that aren't entirely clear, they began to be called "jigsaw puzzles," even though the jigsaw was not used to cut the images to make puzzles.

In any case, the name stuck, and the first cardboard jigsaw puzzles arrived on the scene. These weren't especially popular at first, since they were viewed as lower quality than the puzzles people were used to, which is true. But the innovation did make puzzles more widely available to the public.

One technique that Victorian puzzle makers loved to employ was the fashioning of whimsies, or pieces cut into special, recognizable shapes. By the 1930s, jigsaw puzzles were popular in all levels of society, being fun ways to distract from the economic woes of the age. Though they were still sometimes made with wood, they were by now more often made out of cardboard and came in boxes that had the completed image on the front as a guide. Many "puzzlers" considered consulting this guide to be cheating, so it took a while for the packaging trend to catch on. And increasingly precise saw technology meant that puzzle pieces could be fashioned into new and unusual shapes. These brain-teasers now can be enormous in size, often with thousands of pieces. Spilsbury's simple geography lessons sure have come a long way!

WHO MADE...
CHESS?

One of the world's most popular pastimes, chess is an endlessly fascinating game with an unthinkable number of moves. Indeed, it's estimated that there are more possible variations of the game than there are atoms in the observable universe! (Well, this includes moves not technically allowed in the game, but still). What that means is that no two games of chess will ever be the same, and that it would take several lifetimes to truly master this remarkable game of strategy.

The board layout is deceptively simple and unquestionably elegant. But who devised this amazing pastime? We don't know for sure, but its ancestors can be traced back to at least the seventh century in India, where a game called chaturanga was popular. The word describes a battle formation that was mentioned in the great Indian epic, the Mahabharata. It was a different game, but it had two key ideas in common with modern chess: the pieces carried different powers and values, including a king piece, which was most valuable of all. This version likely evolved from existing board games in the region, changing the rules to create something new.

From there, chaturanga passed into the Arab world, where it evolved into a game called shatranj. With the rise of Islam and increased use of trading routes, the game was brought across North Africa to medieval

Spain. A version of chess may have even been brought back to Scandinavia by Viking traders, where it became the Nordic board game, *hnefatafl*. The famed Isle of Lewis chess pieces, which were found in Scotland and date to the 12th century, also show just how far chess had reached by the Middle Ages.

Chess underwent various other changes and evolutions after the year 1000, both in Europe and in Asia. Despite some religious authorities (both Christian and Muslim) condemning it, it retained its popularity, especially among the upper classes. Attempts to ban it proved futile, and eventually the authorities just gave up, though they retained their negative stance.

The version of chess we know and love today was formulized by the fifteenth century, though not every rule was adopted everywhere, and some rules weren't treated as universal until the eighteenth century. Chess is more popular than ever these days, with clubs, international organizations, and millions of enthusiasts who find it endlessly entertaining and fascinating. There are even a growing number of people who feel that it should be given the status of an Olympic sport—though this gambit is likely to fail.

WHO MADE THE...
ROCKING HORSE?

The beloved rocking horse is viewed as an essential part of Victorian and traditional American childhoods. But this toy has ancestors dating back centuries. Since at least medieval times (and possibly back to ancient Greece and Rome), children have bounded around on stick horses, or "hobby horses," a stick with a representation of a horse's head on one end. Placing them between the legs, the child could run around and pretend to be riding.

By the sixteenth century, some European children were playing with a barrel horse, a log with four legs and a wooden horse's head. While it looked a bit more realistic than the hobby horse, it would have been difficult to move. By the early seventeenth century, bow rockers were invented, and around this time evidence of the first true rocking horse begins to appear. It's speculated that King Charles I of England may have had one as a boy.

Rocking horses continued to be an unusual possession into the eighteenth century. Because of the craftsmanship required, these toys were probably out of reach for all but the wealthiest families. That high cost also means that instead of serving as toys for children, they might have been little more than conversation pieces, allowing the rich to show off.

By the nineteenth century, rocking horses began to take on the form we frequently associate with them, and to become more widely available. Previously, horses had been made of solid wood and so were quite heavy. This, combined with the bow rocker design meant that they could tip over if rocked too hard. In 1880, Philip Marqua of Cincinnati patented a stand to set rocking horses upon, making them less likely to topple over. Manufacturers also hollowed them out, making them more practical and safer.

Queen Victoria was particularly fond of rocking horses, an affection that likely contributed to their popularity in England as the century progressed. Her favorite was the dappled gray design, and this also became the iteration preferred by Victorian families.

After the First and Second World Wars, rocking horses declined in popularity. Or rather, woodworking artisans decided to turn their attention to other projects. Now, sadly, there were very few rocking horse makers left. But the rocking horse remains a symbol of the innocence and rich imagination available only in childhood.

WHO MADE THE...
KITE?

The expression "go fly a kite" tells someone to get lost.
But those who take this admonition to heart are then
thrilled to see this majestic creation defy gravity and
soar through the air. They are fun for children and adults
alike, and in recent decades, advances in design have
produced some truly stunning options.

A cave painting found in Indonesia, dating from 9,500 to
9,000 years ago, seems to depict a kind of kite, known as
a *kaghati*, that is still used by the people in the area. If
this is true, then kites are unthinkably old.

During the fifth century BCE in China, a philosopher
named Mo Tzu is said to have crafted a new kind of
kite, a wooden bird tethered to a line, allowing it to fly
in the air. But was this truly a kite? The debate continues. Around 200 BCE, an account tells of the Chinese
General Han Hsin, who flew a kite over the walls of a
city he was besieging in order to measure how far his
army would have to dig a tunnel underneath them. Not
the lighthearted fun most think of when flying a kite, but
it worked!

Kites in China were used for all sorts of things over the
next several centuries, not just undermining an enemy's defenses. They were employed for testing winds
and weather, measuring distances, communication, and

many other purposes. Also, it's likely some people just thought they were fun! Kite flying spread to other Asian nations, and it became an especially popular hobby in India, where it evolved into the *patang*, or "fighter kite." This version was developed for kite-fighting competitions, where opponents would try to bring each other's kites down.

It's said that Marco Polo brought kites back to Europe from his thirteenth-century travels to China, but kite flying in Europe really took off (see what I did there?) during the sixteenth century, when merchants brought them back from Japan and Indonesia. For some time, they were seen as little more than curiosities, but eventually, scientists began to see value in using them for experiments. Benjamin Franklin's legendary kite-flying venture, where he showed that lightning was indeed electric in nature, is by far the best known of these investigations. Kite experimentation continued throughout the nineteenth century and into the twentieth. The Wright brothers even used kites when working out the design for their first airplane. But these days, they are almost always used for fun and recreation.

WHO MADE...
DOMINOES?

Dominoes are a game in and of themselves, and they have also served other diverting purposes, such as standing in lengthy rows before being spectacularly toppled.

The inventor of dominoes is lost to history. But we do know that the earliest mention of them occurs in China during the thirteenth century; obviously, they could have originated much earlier. A writer named Zhou Mi noted that *pupai* (tiles) were used for gambling. Indeed, the dots on each domino, known as "pips," represent the results of a pair of dice being thrown, which is why each tile has a different combination of dots on it. A rulebook for games using dominoes appeared in the fourteenth century—or maybe it didn't. Some scholars suspect that it is actually a forgery from a later date. In any case, dominoes as an object used in games of chance was well established by the seventeenth century, when more rulebooks were introduced.

The European version of dominoes originates in eighteenth-century Italy. These featured some design modifications (additional tiles and changes in emphasis) and new sets of rules. From there, games featuring the tiles spread to other countries, and the first mention of dominoes as an American pastime occurs during the 1860s. The tiles grew in popularity, taking their place alongside games like checkers, backgammon, and chess.

In the twentieth century, numerous toy and game companies began to mass produce dominoes, with some adding a good number of new tiles (often duplicates). More recently, 55- and 91-tile sets have appeared, allowing for far more possibilities.

But what about the hypnotic toppling we've seen people utilize dominoes for? It's a whole game in and of itself, requiring skill, patience, and vision. It's possible, even probable, that toppling challenges have existed as long as dominoes themselves have. As long as the tiles are made evenly and will stand up straight, anything is possible!

WHO MADE THE...
RUBBER DUCK?

A staple of bath time during childhood, one of those rare, irresistible items that make the excruciating process of getting clean more bearable. As you probably can imagine, the duckie hasn't been around for long, meaning that kids just had to deal with baths beforehand!

The rubber duck made its first appearance in the late 1800s, enabled by Charles Goodyear's invention of vulcanized rubber. These duckies were not meant for the bath, though, as they were solid and didn't float all that well. No, they were intended as chew toys, both for dogs and babies.

Children would have to wait a while longer to relieve their bath-time tedium. In 1931, inventor Eleanor Shannahan produced a hollow rubber duck that could squirt water out of its beak. The idea was to give children something to play with while taking a bath. Then, in 1938, Disney collaborated with Seiberling Latex Products to create floating toys for the bath, including, of course, Donald and Donna (who would later be renamed Daisy) Duck.

In 1947, sculptor Peter Ganine received a patent for his own variety of rubber duck. He painted it bright yellow and added a mechanism to make it squeak. Ganine's duck became an instant classic and began to sell well. It took off even more when, in 1970, the television show

Sesame Street featured a song called "Rubber Duckie," sung by Ernie. The song was so popular that it made the Billboard charts, setting off a new wave of rubber duck love, with sales numbering in the millions!

The penchant for them has never gone away. New versions are constantly appearing, and collectors are keen to snap up rare versions and limited editions. The largest private collection so far stands at more than 5,600 ducks and counting. Who knew such a simple thing could capture the hearts of so many?

WHO MADE THE...
BUBBLE WAND?

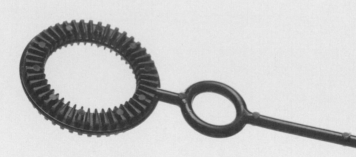

Who doesn't love blowing bubbles? All right, maybe a few grumpy people, but we won't bother with them. From bubble wands to bubble gum, there's something very satisfying about doing little more than exhaling and seeing a bubble grow right in front of your face. If you're lucky, it seals itself and floats away, carried on the breeze. But it just might explode and spritz you with a thin layer of soapy liquid. And, really, that's fine, too!

The idea behind a bubble wand is amazingly simple; it could be made by twisting some wire into a loop and blowing some soapy water through it. So, who invented it? Probably a child or adult looking for something fun to do during a tedious washday! Alternately, bubbles could be been blown out of a pipe or a tube. So, the actual "inventor" of bubble-blowing will never be known.

Whatever its origin, bubble-blowing shows up in artwork at various times. In 1733 or 1734, artist Jean Siméon Chardin painted *Soap Bubbles*, a scene showing a young man blowing a large bubble through a thin pipe, a scene that makes it look very much like glass blowing. In 1886, English artist Sir John Everett Millais painted *Bubbles*, depicting a small child watching a bubble rise from the pipe he's blown it out of.

In 1918, an inventor named J. L Gilchrist received a patent for an improved bubble pipe, that enabled easy cleaning and assembly. Bubble pipes as toys were popular for the next several decades and often came in sets with soap and a dish for the soapy water.

As for the bubble wand? In 1960, inventor Arthur Fulton created a flexible loop that could be inserted into a jar of "bubble water" and would expand when drawn out, creating much larger bubbles without the need to blow on them. In 1987, David Stein improved on this design and was granted a patent for a wand that was more stable and able to create even bigger bubbles. Stein's innovation ignited a new bubble-making craze and created a new competitive sport, as people vied against each other for who could make the largest, longest in length, and longest-lasting bubbles.

Since then, practitioners of the art of bubble making have produced some astonishing results, including bubbles that are over 100 feet long!

WHO MADE THE...
PUPPET?

Puppet can mean many things, from little figures perched on fingers to marionettes on strings, from the uber-creepy ventriloquist dummy to people willingly manipulated by others. This entry focuses on the playful versions operated by hands and strings, in case you were wondering. Puppets and puppetry have a long tradition around the world, and may have arisen simultaneously in several cultures. After all, using puppets to tell a story is not a big leap from the common occurrence of a child giving a toy a personality and voice of its own.

Puppets seem to have been originated about 4,000 to 4,500 years ago, as archaeologists found a clay doll with a head that was moveable by a string dating to this period in the Indus Valley. This could be some form of early puppet, or it might be something else altogether. A monkey that could be animated by a stick was also discovered during this dig, an object that seems to lie somewhere between a toy and a proper puppet.

Puppets could have played roles in rituals in ancient Egypt, as early as 2000 BCE. There is a description in one Egyptian text of "walking statues" being used in religious dramas, which could be some kind of puppet.

China has a long tradition of puppetry, and the earliest surviving mentions of it come as early as 1000 BCE. Chinese puppet shows were mainly shadow theater, where the puppets are manipulated by sticks behind a thin screen, lit from behind. The shadows of the puppets would then be moved by the hidden operator, making them seem life-like. This technique found favor in India and Southeast Asia, where shadow theater is still a popular traditional form of entertainment.

Puppetry was known in ancient Greece and Rome, as the figures were used in storytelling and theater. This tradition continued in Europe into the Middle Ages and only grew in popularity over time. Despite disapproval from the Catholic Church (per usual), puppets were often used in morality plays and depictions of biblical stories to present these tales to a largely illiterate audience. Indeed, in Italy during the Renaissance, the Church made use of marionettes to tell religious stories, and it's thought that the very word *marionette* comes from the miniature statues of the Virgin Mary.

Puppet theater only grew in popularity over the next few centuries, with marionette shows evolving into full-scale productions that established its practitioners as artists. After peaking with the "Punch and Judy" shows that were a popular attraction in British seaside towns, puppetry has taken a back seat to other entertainments—but its ability to charm people of all ages has never dwindled.

WHO MADE THE...
ALARM CLOCK?

Some (most?) just like to sleep in. Others are intent on forcing them to get up, which is, honestly, very inconsiderate. But because this unpleasant imbalance exists, humans have been trying for centuries to create machines that will help them wake up earlier than they would of their own volition. The history of clocks that can rouse people at horrific hours is far longer than you might think, which means that sleep deprivation is far from a modern phenomenon.

The famous Greek philosopher Plato is said to have possessed a water clock that would sound an alarm, possibly to let him know when it was time to go and give lectures. The sixth-century Roman statesman Cassiodorus urged monasteries to use water clocks to rouse monks to their duties. And around the year 725 in China, Buddhist monk and inventor Yi Xing created a mechanical clock that could sound alarms. He called it the "Water-Driven Spherical Bird's-Eye-View Map of the Heaven," which, you have to admit, is a lot more lyrical than alarm clock.

In 1319, the Italian poet Dante Alighieri described a mechanical clock that could be set to strike at three different times each day, and in Europe, clocks that could be set to chime at certain hours date back to the fifteenth century. These clocks tended to be public devices, or for use by religious houses and the like.

In 1787 in Connecticut, clockmaker Levi Hutchins crafted an alarm clock for his own personal use; he set it to wake him up at 4:00 a.m. each morning for work, though he didn't seem to want to spread the misery around and kept the device to himself. In 1847, French inventor Antoine Redier became the first to patent a programmable alarm clock, and then the terror of a too-early morning was inflicted on people across the Atlantic when American inventor Seth E. Thomas created and patented his own version in 1876. Thomas's company made alarm clocks affordable enough for the masses, so yeah, blame him the next morning you wake up bleary-eyed and in a panic.

Adding insult to injury, inventor James F. Reynolds came up with the idea of the clock radio, an alarm that could wake you up with a tinny-sounding radio broadcast every morning. Yipee for him. Today, quite a few people use their phones as alarms, as they use their phones for pretty much everything. There are even "gentle alarms" which can wake you with a series of chimes that gradually increase the volume to nudge you from sleep, rather than shock you out of it by a heart attack–inducing shriek. But, no matter what spin inventors try to put on it, the alarm clock remains an enemy of the people.

WHO MADE THE...
ERASER?

The simple act of using a bit of rubber to wipe out a mistake takes a lot of pressure off, whether in taking a test, drawing a picture, or doing a crossword. Erasers fill such an obvious need that it seems odd that they haven't existed for as long as writing implements have.

Prior to rubber erasers bouncing into our lives, other materials were used to remove pencil marks. Wax could remove some markings from some surfaces, and pumice stones could rub out both pencil and ink, though these stones also wore away the surface of the paper, and so had to be used sparingly. One of the more interesting early erasers was bread. During the second half of the nineteenth century in Japan, people would remove the crust from a piece of bread, moisten what remained, and fashion it into a ball. These balls were then used to blot out mistakes made in ink or pencil. One student recalled how they were given as much bread as they wanted for erasing, but some students took extra and ate it when they were hungry. We can presume (hope?) that they didn't also consume the lead- and ink-stained bread used to keep their work tidy!

But relying on pumice and bread wasn't going to cut it. In 1770, an English engineer named Edward Nairne accidentally discovered the erasable properties of rubber when he reached for some bread crumbs to undo some pencil markings he'd made, but instead picked up a piece of rubber. Nairne was astonished that the rubber took the pencil right off, with no damage to the paper. He knew he was onto something! Nairne began to sell chunks of rubber for this purpose, and the idea caught on quickly.

In fact, the term *rubber* comes from the substance being used to "rub out" markings on paper. The name has ever since been used in Britain, while Americans came to prefer the term "eraser." Rubber as a material is perishable, however, a problem until Charles Goodyear (whom we've already met, see page 220), and his vulcanization process came along in 1839. In the 1850s, a man named Hymen Lipman came up with the ideas of attaching a miniature eraser to the end of a pencil, and the rest is writing (and drawing, and doodling, and calculating) history. The pencil with eraser is now such an established part of our lives that it's difficult to imagine the world without it!

WHO MADE...
TOILET
PAPER?

If a creature eats, it's very likely it excretes. This has been true for hundreds of millions of years, across all kinds of species. While a comprehensive history of pooping is way beyond the boundaries of this book, suffice to say: humans have long history of cleaning up after themselves once they've finished their business. It's just good hygiene and the alternative is very unpleasant. We all use toilet paper of some sort these days, and its importance could be seen in the ridiculous mobs that tried to hoard it during the early days of the COVID-19 pandemic.

But where did this attachment originate?

The use of actual paper for cleanup purposes comes from China. There are multiple references to the practice going back to the sixth century and continuing over the next several hundred years. By the fourteenth century, millions of small packages of paper for wiping purposes were being produced in various Chinese provinces, most for use by the upper classes. Some of these varieties were even perfumed!

In Europe, the idea of toilet paper caught on much later, when paper became cheap to manufacture and thus widely available. Even then, paper specifically for sanitary purposes doesn't seem to have been a thing. Instead, it was recycled from newspapers and old books.

In England in 1747, a certain Lord Chesterfield recommended the use of paper for cleaning up, telling his son of a man who, "bought [...] a common edition of Horace, of which he tore off gradually a couple of pages, carried them with him to that necessary place, read them first, and then sent them down as a sacrifice to Cloacina." Cloacina, by the way, was the ancient Roman goddess of the drains and sewers of Rome. All in all, this seems like a pretty good system: spend some time with the classics, clean up, make good with the old gods, and go happily on one's way.

What we think of as toilet paper came along in 1857, when American inventor Joseph Gayetty introduced Gayetty's Medicated Paper, which was sold in packages of sheets. The first rolls of toilet paper were sold by the Scott Paper Company, though designs for rolled paper had been patented some years earlier (toilet paper seems to be another one of those ideas whose time had come). The increase in the number of homes with flushing toilets by the end of the nineteenth century only made it that much more popular.

TP designs continued to improve, though apparently not as quickly as some would have preferred. As recently as 1935, the Northern Tissue Company proudly boasted that its toilet tissue was "splinter free," which speaks to quite a prickly past! But gradually, toilet tissue did become softer, and then along came two-ply, and all the other innovations we've come to know and take for granted— so long as supplies don't run out.

WHO MADE THE...
RUBBER BAND?

They're essential for keeping things bound together and, most important of all, for providing distraction in the form of firing across the room. The rubber band is a staple at the office and around the house, but who came up with this ingenious idea?

We've already talked about how Charles Goodyear's vulcanization process made rubber durable and water-proof, which was important for everything from erasers to automobile tires. And it was important for the humble rubber band, too.

A British inventor named Stephen Perry saw the value in using vulcanized rubber for a variety of tasks. He founded the wonderfully named Messers Perry and Co, Rubber Manufacturers of London, and in March 1845, he patented the world's first rubber band. It was intended to hold together stacks of paper, and it did so just fine. These bands were timely, as the rapid growth of industry and business in Victorian England was generating a lot of paperwork.

Perry's idea wasn't new, of course. People had been tying together everything from paper to sticks with strips of cloth or leather for thousands of years. The Mayan people had even used a kind of latex from rubber trees to make a variety of products, and it seems almost certain that among these were cords for trying things together.

But Perry's vulcanized rubber bands had the advantage of being very durable; they wouldn't snap unless stretched way too far. Also, it would take a considerable amount of uses before they became brittle. Thus, his rubber bands were practical and long-lasting. It wasn't too long before other inventors and industrialists began to see the value of these bands for machines and tools, and the market for rubber bands expanded dramatically.

Even in today's highly digital world, we still rely on them. But let's be honest, shooting them across a room is still their highest use!

WHO MADE THE...
SYRINGE?

We most often associate the syringe with the hypodermic needle, and it's true that this is a very common form of the device. But syringes existed long before needles haunted our nightmares, and they didn't always have medical uses. The word itself comes from the Greek syrinx, meaning "tube."

In the first century, Roman encyclopedist Aulus Cornelius Celsus described an early form of the syringe in his book *De Medicina*. Often these tubes were used for anointing people with various oils, either for medical or ritual purposes. In the second century, the physician Galen described similar devices. By the tenth century, ʻAmmār ibn ʻAlī Mawṣilī, an Arab doctor, performed successful cataract surgeries by using a hypodermic tube to remove the cataracts via suction. Yes, it sounds horrible, but it worked, restoring sight to many patients.

By 1650, mathematician, philosopher, and inventor Blaise Pascal created what might be considered the first hypodermic needle, and the architect Christopher Wren also dabbled in the technology. The problem was managing dosages and sterilizing needles, which pretty much didn't happen. Which meant that the patients died. Which meant using these kinds of needles quickly fell out of favor and wasn't reconsidered for quite some time.

Finally, in 1844, Irish physician Francis Rynd created a hollow needle, capable of safely administering subdermal injections. This was followed in 1853 by Charles Pravaz in France and Alexander Wood in Edinburgh independently developing a syringe-hollow needle combination that allowed for the accurate measurement of doses. The basic design of these tools hasn't changed all that much since. Obviously, syringes have been tweaked, and many now use plastic rather than glass, but the idea, and its significant life-saving potential, remains the same.

WHO MADE...
WI-FI?

Trust me: Wi-Fi is everywhere, even if you can't see it. Younger people can't imagine a time when it was not widely in use, and even some older folks are thrilled by its existence. Unlike many inventions in this book, it doesn't have a long history. But the idea does go back further than you might think, to 1971. Under the direction of a computer scientist named Norman Abramson, a networking system called ALOHAnet was connected in the Hawaiian Islands. It used UHF, a forerunner of wireless technology and ethernet.

In 1985, the Federal Communications Commission in the United States released these frequency bands for free use, meaning they didn't have to be licensed. Companies in Australia and the Netherlands began to experiment with wireless technologies. The NCR Corporation, working with researcher Vic Hayes, developed what are now known as the IEEE 802.11 standards for how the system should work. As such, Hayes is often considered the "father of Wi-Fi." The standards were released in 1997, and updated two years later. Although slow by today's standards, its release was a major step forward in bringing wireless technology to the public.

One of the earliest adopters was Apple, which included it as a technology in the iBook laptops in 1999. Once such a major company gave it the green light, other computer companies and even whole industries began to come aboard. The technology continued to improve, and is becoming faster and more reliable. While there are still some advantages to being "plugged in," these days, most internet and communication technologies rely on some kind of wireless setup, a dependence that is unlikely to diminish any time soon.

WHO MADE THE...
MIRROR?

The most basic mirror is the shimmering image on the surface in a pool of water. As such, creatures of all kinds have been gazing at themselves for untold millions of years. Whether or not they recognized themselves is, of course, a whole other conversation. But what about actual mirrors, constructed for the purpose of vanity? Or at least making sure no food is stuck to one's teeth?

Evidence of mirrors made from obsidian has been found in Anatolia, what is now modern Turkey. These date back an amazing 8,000 years and show that even then, personal appearance and grooming were of great importance, at least to a certain class of people. By 6,000 years ago, mirrors of polished copper were known in Mesopotamia, and soon after in Egypt. From about 4,000 years ago, we have evidence of mirrors made of highly polished stone in Central and South America. Mirrors made of bronze were known to people in Nubia and China from about 2,000 BCE.

Mirrors containing glass were present in the Roman Empire from the first century CE, and these continued to be improved on well into the Middle Ages and Renaissance. The mirror as we know it today was an innovation of a German chemist named Justus von Liebig. In 1835, he came up with the idea of adding a thin layer of metallic silver to a sheet of clear glass. The result was a very clear and reliable mirror. This technique was easy to manufacture, and it meant that mirrors could be sold to everyone, greatly increasing their popularity. This technique also gave people a very clear image of themselves, whether they liked it or not!

Over the decades, there have been many improvements in the initial design, such as adding coatings to protect from abrasions, but the basic design isn't all that different from von Liebig's, whether for everyday or industrial use. As for whether breaking one brings bad luck? That's still up for discussion, but it's best to be extremely careful, just in case!

WHO MADE...
VELCRO?

Velcro is an ingenious invention, securing all kinds of items, from pants and backpacks to shoes. The principle is simple enough: thousands of tiny hooks grab on to equally tiny loops and hold fast. It brings to mind one of those annoying burs that tend to attach themselves to your socks and pants when you're out hiking, and are very difficult to dislodge. In fact ...

In 1941, Swiss inventor George de Mestral was taking his dog for a walk in the Alps. After some time, he noticed tiny burdock seeds clinging to his socks and pants, and also to his dog's coat, seeds that could not be easily shaken off. Curious, de Mestral decided to examine them under a microscope and was surprised to find that tiny hooks in the seeds were holding on to the loops in his socks and his dog's fur. It got him thinking: What if he could replicate this pattern artificially to create a substance that could be used to fasten things?

It would be a powerful alternative to buttons, laces, and zippers, and especially useful for those, say the elderly, who might have trouble fiddling with these items.

Initially, as is often the case with such breakthroughs, no one in the industry was interested, with de Mestral's idea even being ridiculed. Undeterred, he experimented with natural fibers but found that they frayed and wore out too easily, so he began working with the more robust synthetics available to him. Eventually, he settled on using nylon, then tinkered with the design over many years to perfect it. Finally, he was granted a patent for his invention in 1955, first in Switzerland, and then in many other European countries and the United States.

De Mestral came up with the brand name "Velcro," a portmanteau of the French words *velours* ("velvet") and *croché* ("hook"). It was a simple, easy-to-remember name. Initially designers and clothing manufacturers were hesitant to use it, but after the aerospace industry adopted it for various items, its profile was significantly improved. Skiers soon found that the material worked very well for certain gear, which further enhanced Velcro's image. People began to see it as a futuristic innovation, and Velcro was soon showing up on jackets, shoes, and just about anything that needed to be fastened.

After de Mestral's patent expired in the late 1970s, several other companies jumped on the bandwagon and started creating their own versions. Though it's no longer seen as the fastener of the future, Velcro and its imitators are still keeping much of the world held together.

WHO MADE THE...
NEON SIGN?

Once upon a time, they seemed to be everywhere. And even now, neon signs still point the way to businesses, pizza places, dive bars, strip clubs, and anything else that needs a little help standing out in the dark of night. People can get quite creative with these colorful signs, and many are works of art themselves.

Their story begins in 1855, when inventor Heinrich Geissler created the proudly named Geissler tube. When the tube was placed under low pressure and electrical voltage was added, the gas inside began to glow. Various inventors played around with this for several decades, trying to find a practical application. In 1898, British scientists William Ramsay and Morris W. Travers discovered neon gas, a rare element in Earth's atmosphere, but one which lit up spectacularly within a Geissler tube.

In 1902, French inventor Georges Claude created a company that produced neon gas as a by-product. Claude came up with the idea of using electrically charged neon to make a lamp, and displayed it in 1910 at the Paris Motor Show. One of his associates saw potential to create signs that lit up the night, and so these signs began to appear over the next several years, including lighting the Paris Opera by 1919, a major achievement for the time!

Claude patented various innovations to his lamp and brought his company, Claude Neon, to the United States, where he successfully sold neon signage to a number of big companies. Neon naturally glows red, but in the 1920s inventor Jacques Risler devised the idea of fluorescent tube coatings, which could be used to change the color. Initially there were only a few dozen, but soon the color palette expanded, and now there are more than 100 options.

The heyday of neon signs was the 1940s through the early 1960s, after which their popularity began to wane. After a brief revival in the 1980s, neon signs again declined, and today LED lighting is much more commonly used. Though it is interesting that, on occasion, LED lighting is sometimes made to look like classic neon signs, scratching our itch for the unique aesthetic they provide.

WHO MADE...
LIPSTICK?

One of the most popular of all cosmetics, lipstick has very ancient origins, so it's impossible to identify a single inventor. Evidence points to the first regular use of coloring for the lips (for both men and women) occurring in both the Mesopotamian and Indus Valley civilizations, over 5,000 years ago. In both cultures, using color on the lips was probably a way to signal high social status. The Mesopotamians used a mixture of gemstones mixed with ointments for lipstick, so it was likely very expensive, not something for the commoner. The lip coatings in the Indus Valley seems to have been made from ochre.

From there, the tradition of covering the lips never faded (unlike lipstick itself!) and was adopted by the Egyptians, again used by both men and women. In Egypt, it was a status symbol but was also used to protect against the harsh sun. In some civilizations, such as the Minoan, lip coloring was initially reserved for prostitutes (albeit those of elevated status), but it ended up being widely adopted by the upper classes as well.

At least 1,000 years ago, Chinese women wore lip coatings made from beeswax to protect them from the elements, though these balms may have been more like the wax sticks used by sunbathers and skiiers today. In Muslim Andalusia, the tenth-century physician and chemist Abu al-Qasim al-Zahrawi studied scented cosmetics and considered them a crucial part of overall health. He invented a perfume that could be pressed into a mold and then applied. Adding color to this, al-Zahrawi created what is considered the first lipstick in the modern sense of the term.

In Europe, lipstick went in and out of fashion over the next several centuries. Initially condemned by religious authorities as sinful (it had been worn by pagans and barbarians, after all), it gradually came to be accepted, or at least tolerated (or the Church could no longer do anything to stop it). Once again, the upper classes led the way. In Tudor England, Queen Elizabeth was quite fond of "lip rouge" and used her own, specially made formulas. By the eighteenth century, upper-class men were once again just as enthusiastic about wearing it as women.

But during the Victorian Age, things changed. After the death of her beloved husband, Albert, the queen imposed a ban on lipstick everywhere in the British Empire, but secretly and gradually, it started to come back into use. Lipstick came back into fashion with the women's rights and suffragette movements, and major cosmetics companies were eager to get in on the craze. With the rise of Hollywood movies and movie-star glamour, lipstick became permanently situated in everyday culture, though mainly for women only; men have yet to regain the cultural "permission" to wear anything other than clear balms on their lips.

WHO MADE THE...
ZIPPER?

Quite simply, zippers keep things held together: clothes, bags, luggage, shoes, and so much more. Without them we'd be spending extra hours each year buttoning and unbuttoning. An ingenious invention that isn't obvious right off the bat, the zipper wasn't just waiting for someone to come along and "discover" it. So how did it come about?

The potential ancestor of the zipper dates back to 1851, when an American inventor named Elias Howe (who also invented the modern lockstitch sewing machine) received a patent for his "Automatic, Continuous Clothing Closure." That's a mouthful! Howe was trying to create a method of fastening clothes that went beyond buttons, but for whatever reason, he didn't really pursue it, even after going to the trouble of getting it patented. Maybe he was so preoccupied with the overwhelming success of his sewing machine that he didn't have time?

Four decades later, in 1893, another American inventor named Whitcomb Judson developed Howe's idea and came up with the "Clasp Locker." Many consider him to be the true father of the zipper, but his invention failed to make a splash, even when he exhibited it at the World's Fair in Chicago that year. In any case, though Judson's invention was not called a "zipper," designs for it do resemble modern zippers, and he launched a company, Universal Fastener Company, to sell it.

The company hired a Swedish engineer, Gideon Sundback, who developed the design into something much more useful and graceful by 1913. This "Separable Fastener" was patented in 1917. After this, the design began to take off, as companies began to see its usefulness in a variety of ways—though not really for clothing at first.

The name came from the B. F. Goodrich Company, who used the fasteners beginning in 1923 for a brand of rubber boots. Another popular use for the zipper was tobacco pouches, of all things. Zippers finally appeared in children's clothing in the 1930s, but it took a while for adult clothing to catch up. In 1937, French fashion designers came around to the idea of using a zipper on the men's trouser fly (of course it would be the French!), and suddenly, they were respectable.

Since then, zippers have spread their effortless nature everywhere, with estimates claiming that several thousand miles of zippers are produced daily to meet the demand.

WHO MADE THE...
GARDEN GNOME?

The humble garden gnome has a longer history than you might expect. Love them or loathe them, these little ladies and fellows have been adorning gardens in Europe and the United States for over 150 years, but their ancestors go back much further in time, to ancient Rome.

The Romans were convinced that the supernatural surrounded them at all times. While there were the "big" gods at the temples, such as Jupiter, Neptune, Mars, and so on, people didn't have much interaction with them on a daily basis; they were very busy gods, after all. No, they had their own personal house spirits and certain protector gods that were not as important in the pantheon, but still presided over certain things, such as keeping homes safe. In addition to personal house spirits, statues of the god Priapus were popular, since he could provide a home and garden his blessings and protection.

Priapus was always depicted as a short, squat fellow (rather like a garden gnome) with one, um, notable exception: an enormous phallus, which could be as long as he was tall. He was often placed in a garden to ensure protection and fertility.

The word "gnome" itself probably comes from the Greek *genomos*, or "earth-dweller." But the idea of these small, earthbound spirits was not limited to the Greco-Roman world. In the Germanic and Scandinavian regions, there were myths about all kinds of spirits: dwarves, gnomes, goblins, and so on.

And it was in these countries that using little creatures to adorn the garden came back into fashion fully in the eighteenth and nineteenth centuries. Various German and Swiss makers began producing little figurines for homes and gardens. In 1847, Sir Charles Isham, a keen English gardener, brought back 21 terra-cotta figures from Nuremberg, Germany. He named them "gnomes" and proceeded to set them up in gardens. Then, as now, some people loved them, and others, well, not so much, including Isham's own daughters.

The little figurines weren't an overnight sensation. Many people saw them as kind of tacky and unworthy of decorating a "serious" garden. But that didn't stop industrious German factories from cranking them out in large quantities during the late nineteenth century. Interest in garden gnomes and things Germanic waned during and after the World Wars, but was revived in later on. It's thought that today's garden gnome designs are partially based on Disney's *Snow White and the Seven Dwarfs*, from 1937.

Garden gnomes have never gone out of style, though they

haven't exactly ever been in style, either. Like them or not, they are curious part of gardening lore, with roots stretching back to a very well-endowed Roman god.

WHO MADE THE...
LAWN
FLAMINGO?

Like the garden gnome, this odd lawn adornment is something you either love or disdain. Most often the latter. But that doesn't stop people from sticking them in their yards. It all seems a bit odd. Statues of flamboyant birds in the yard? Maybe, but why flamingos? Why not parakeets, or blue-footed boobies?

Contrary to popular belief, these odd lawn ornaments didn't originate in Florida, but in a much more unlikely place: Leominster, Massachusetts, the self-proclaimed "Plastics Capital of the World." In 1957, a young sculptor named Don Featherstone was hired by a plastics company called Union Products to create a plastic flamingo. The motivation for this project is now forgotten, but there you have it.

The only problem Featherstone had was a lack of access to live flamingos. The weather in Massachusetts is not exactly flamingo-friendly, you see. Instead, he sought out some *National Geographic* photographs and created his designs based on those. Probably no one thought they would be anything other than a curiosity, but for whatever reason, they began to take off, just not literally. Perhaps it was the sameness of 1950s suburbia that did it: people wanted something that made their lawns stand out, and maybe said flamingos were the right measure of kitsch and colorful.

The pink birds became a bit of craze in the 1950s and early 1960s, but soon they began to fade, as social unrest made some see the plastic flamingo as nothing more than a symbol of the cheap and tacky present in American life, and of the wasteful plastic industry.

They became an avatar for lack of taste, to be derided or displayed ironically. Still, they never completely went out of fashion, and after changing ownership and rights a few times, these odd little relics of the 1950s are still being produced today.

Fun fact: In 2009, the city of Madison, Wisconsin, probably with tongue planted firmly in cheek, designated the plastic flamingo as its official bird.

WHO MADE THE...
LAVA LAMP?

They're fun, weird, and were all the rage back in the late 1960s and early '70s. They've made a comeback as a retro-cool item in recent decades, even though they're not very good at being lamps, to be honest. We're talking about lava lamps, of course! Yes, the thing that no dorm room or neo-hippie hangout can do without.

This odd little bastion of the counterculture got its start in the early 1960s. British accountant Edward Craven Walker (who, for the record, also made underwater nudist films) saw a homemade egg timer made out of a cocktail shaker on a stovetop. The liquid inside was bubbling, and it gave him an idea. Okay, it was an odd idea, but it would prove to be very fruitful. Working with fellow British inventor David George Smith, Walker suspended globs of wax in liquid that would heated up by a light source. Calling it the "Astro," they devised various sizes and shapes.

Their invention didn't really go anywhere until 1965, when two men bought the American rights at a German trade show and began to market it as the "Lava Lite Lamp." They weren't intended to be symbols of the resistance. Indeed, there was even an executive model that was advertised in a 1968 issue of the *American Bar Association Journal*, which kind of boggles the mind!

But for whatever reason, these odd, alien-looking lamps were embraced by a generation of young protestors and became must-have items. And so it was that the lava lamp became a the height of cool. But, like all fads, they began to go out fashion as quickly as they became essential, and ownership of the company changed hands several times over the next few decades. By the end of the 1980s, they were little more than historical curiosities. But then something unusual happened: in the 1990s, the *Austin Powers* movies ushered in an unprecedented nostalgia for the 1960s. And with that came all sorts of accessories, including lava lamps.

Suddenly they were hip and cool and again, and sales started going through the roof. It's estimated that a million or more now sell every year, something that Edward Craven Walker probably never envisioned when he encountered that egg timer!

WHO MADE THE...
BALLOON?

They're a favorite of children, especially the ones filled with helium that can float on their own. And filling balloons with one's own air, only to let them go and watch them zip across the room as they issue fart-like noises is another youthful pleasure. Of course, many a child has been crushed with disappointment when they let go and watch their balloon defy gravity and sail away to other adventures.

The idea of filling a container with air so that it could float is quite old. It's thought that the Aztecs invented balloons out of animal intestines and might have used them to make sculptures and as sacrifices to their gods. The Greeks and Romans also might have participated in this practice, and the people in the medieval period almost certainly did it. In the seventeenth century, Galileo made use of an inflated pig's bladder when he conducted an experiment to measure the weight of air.

Large balloons became a craze in the late eighteenth century, when the first experiments in hot air balloons were carried out to some success. On a smaller scale, rubber balloons similar to what we know now were invented by the English scientist Michael Faraday, who wanted to use them in his experiments. By 1847, J. G. Ingram began manufacturing balloons made from vulcanized rubber (which, as we've seen, was revolutionizing industry at the time), and these were the direct precursors of modern toy balloons.

In the 1920s and '30s, Neil Tillotson invented both latex balloons and latex gloves. He created a latex balloon featuring a cat's face and ears, with the obvious intent of entertaining children, and managed to sell bundles of them. By then, it seemed to have dawned on folks that balloons had entertainment value. As seen by their presence at birthday parties and many other celebrations, and their ability to put a smile on most everyone's face, that value is far from being exhausted.

WHO MADE...
HIGH HEELS?

High heels are both loved and reviled by those who wear them. Occasionally, the heel is low and practical, but often, it's ridiculously high and uncomfortable. And although heels are usually associated with women's fashion, they had a very practical origin, and were worn by men.

In tenth-century Iran, some unknown genius discovered that if soldiers had a low-heeled boot, they could more effectively rest their feet in stirrups, allowing the rider to stay safely on the horse's back and direct the animal better. This gave mounted cavalry an advantage in attack and horsemanship. While this particular innovation stayed in the region for several centuries, heeled shoes were worn by European nobility, as they wanted to be kept as far away as possible from any filth in the streets they walked down.

The high-heeled shoe as we know it probably came into Europe via trade with Iran in the sixteenth or seventeenth centuries. They soon became a symbol of status, with their obvious impracticality for daily life showing that the wearer didn't have to work. Again, they were worn mainly by men at first, though by the latter half of the seventeenth century, women had their own designs being made. Men's heels became thicker, while women's became thinner as a way of distinguishing them. Neither were particularly high (though those worn by women became higher throughout the eighteenth century), but they were offset enough to show that the wearer was involved strictly in pleasurable activities. But by the 1790s, at least in France, heels came to be seen as symbols of the hated nobility, went out of fashion, and remained so for a time.

In the second half of the nineteenth century, high heels began to make a comeback, but only for women. There were a few exceptions to this, such as the popularity of cowboy boots in the American West, but here, just as in tenth-century Iran, the heels were low and served a purpose; they were thick, heavy, and made it easier for the feet to remain in stirrups.

Women's heel fashions continued to evolve for several decades, reaching extremes in the 1950s with the arrival of the stiletto. By the 1960s and 70s, many women were rejecting heels as a product designed for the male gaze, but they came back into fashion again in the 1980s, and many women (and men) now wear them proudly. What started as a practical invention for war and became a way to flaunt wealth has become an essential fashion accessory, even if it's loathed by a lot of people!

WHO MADE...

SILLY STRING?

It's string; it's silly. What more do you need to know? Well, a lot, actually, because the origins of this whacky toy are very different from what you might imagine. Silly String wasn't supposed to be a toy at all. In fact, it was created for medical reasons. Wait, what? Yes, it's true, the product that would become Silly String was originally not silly in the slightest.

In 1972, inventor Leonard A. Fish and chemist Robert P. Cox were working on developing a product to treat broken bones. The idea was to create a sticky spray-on substance that could be applied to a broken limb. It would then harden quickly and create a makeshift cast. This would be invaluable to people working in remote areas, hikers, laborers of any kind, the military, etc. Just think: a portable, moldable cast in an aerosol can!

It was a great idea, but when the pair started working on ways to actually spray it, the results were less than great. Fish tested a large number of nozzles, and while working with one, he shot the substance across the room, some thirty feet. Almost immediately, it struck him that the product could be a great toy. After working to make the substance less sticky and more, well, stringy, Fish and Cox decided that they would need help in marketing it, so they contacted Wham-O and made an appointment.

According to Fish, when they went to demonstrate the product, he ended up spraying it all over the hapless rep. The man, not at all pleased, told them to leave at once, and they did, feeling pretty dejected. But a day later, Fish received a telegram from Western Union. It was from the man he'd inadvertently sprayed, asking for samples to be sent over. The owners of Wham-O had apparently heard about the incident, seen some of the material, and wanted to examine it further.

A couple of weeks later, Fish and Cox agreed to license the product to Wham-O under the name "Silly String," and the rest is history. Imagine what would have happened if the angry rep had never been made to contact them? How much silliness might the world have been deprived of?

WHO MADE THE...
CORKSCREW?

It's an ingenious idea, and one perfectly suited to removing the cork from any bottle, particularly those filled with wine. In fact, it can be argued that excising the cork with a corkscrew is a large part of the ritual of drinking and enjoying wine. But like so many inventions, the corkscrew has its origins in something entirely unrelated, in this case, guns.

What?

Yes, the earliest mention of a corkscrewy entity was the "gun worm," an implement whose use was first noted in England during the 1630s. It was used to clean out the barrels of muskets and pistols, clearing away unspent bullets and any debris that had become wadded up.

At this time, glass wine bottles of various sizes were coming into regular use, many of them sealed with corks. The problem: there was no tool that could extract those corks without making a mess or shattering the glass.

It's thought that the corkscrew was invented in England, and the earliest reference to it comes from a mention in 1681 describing a "worm" that can pull corks out of bottles. Undoubtedly, there were many forms, but the first patent for a corkscrew was and granted to the Reverend Samuel Henshall in 1795.

Many types of corkscrews appeared over the next century, as inventors kept trying to come up with better and easier ways to get to that all-important wine without much struggle. A new design, the wing corkscrew (first patented in Britain in the 1880s and in America in the 1930s), made cork extraction far easier, as flipping up the wings once the corkscrew was in did all the work; no stressing and straining to remove a stubborn cork that might break when it was halfway out! The famed "sommelier's knife" arrived in 1882, and has been a preferred tool of wine specialists ever since. This tool, which features a small knife and a metal fulcrum in addition to the corkscrew, is also known as the "waiter's friend" and the "wine key."

Fancier ways of extracting corks, such as the lever corkscrew, came along in time. The twin-prong cork puller inserts those prongs on either side of the cork and pulls it out. This has the advantage of not damaging the cork, should you wish to replace it back into the bottle. The lever corkscrew mechanism allows for very easy cork extraction, but is heavier and bulkier, and, for many people, not worth the extra effort required.

In the end, people have always chosen whichever corkscrew feels best. There's no right or wrong version to use, just make sure you choose one—if you don't, you'll be stuck online, watching videos on how to remove a cork with your shoe!

WHO MADE...
BUBBLE WRAP?

We love bubble wrap for keeping things safe when packed up, and we love it even more once we can pop the bubbles! The feeling is so pleasurable that there are now websites where you can pop virtual bubbles, complete with the appropriate sound—and the best thing is that you have an endless supply!

But what's the story behind this ingenious invention? Surely, it must have been created to ensure that things could be shipped safely, right? Well, no. It actually started out as a form of wallpaper.

In 1957, engineer Alfred Fielding and his friend, the chemist Marc Chavannes, were working to create a kind of textured wallpaper. At some point in the process, they took two pieces of plastic shower curtain, put them together, and ran them through a heat sealer. They weren't especially pleased with the result: two sheets of plastic fused together, with some air bubbles trapped in between them. But they felt they were onto something and kept working at it. As there didn't seem to be a market for plastic wallpaper with bubbles, Fielding and Chavannes tried marketing their creation as greenhouse insulation, but again, the interest just wasn't there.

In 1961, IBM needed a safer way to ship its very delicate and valuable computers, and various heads were put together to come up with a solution. It turned out that plastic sheeting with bubbles between the layers was the perfect way to wrap IBM's electronics and ship them safely. Fielding and Chavannes tinkered with their creation until it was suitable as a packing material.

As it turned out, bubble wrap was far better than the alternative materials available at the time, paper or newspaper. It offered far more insulation, and (unlike newspaper) didn't get ink all over the item. The invention had found its calling, becoming a hit with numerous companies that needed to ship packages to customers. Fielding and Chavannes experimented with different sizes of sheets and bubbles, creating different designs depending on what a particular customer required.

Regarding the other important ability of bubble wrap? Well, Fielding's son, Howard, believes he might be the first to have realized the pleasure of popping bubbles. At the age of five, he had a sheet of an early prototype in his hands, and popped one of the bubbles, and then another, and another. He notes: "The bubbles were a lot bigger then, so they made a loud noise." And so, a pleasurable pastime was born!

WHO MADE THE...
NECKTIE?

The necktie is, for better or worse, still considered an essential part of men's business attire and formal wear. While they can look smart and classy, they can also be very uncomfortable, and some studies have suggested that wearing them every day can have negative long-term effects on one's health, with issues related to circulation and blood vessels potentially arising. Despite that grim prospect, neckties have ensnared men all over the world.

For some time in Europe, upper-class men and women wore decorative pieces of cloth around their necks. By the sixteenth and seventeenth centuries, this tendency reached a bizarre extreme in the ruff, a large piece of circular, starched fabric worn like a collar. They were highly impractical and became increasingly more so before the fashion faded after nearly a century.

But the upper classes were not done ornamenting their necks. It's believed that during the Thirty Years' War (1618 to 1648), King Louis XIII of France hired a troop of Croatian mercenaries who wore a certain kind of tie around their necks. They called themselves the *Hrvati*, while the French referred to them as the *Croates*. It's thought that the word *cravat* came from these two words, providing a name for these elaborate ties. The young King Louis XIV took a liking to the cravat and began wearing them in the 1640s. Of course, whatever the king did set off a fashion trend, so extravagant cravats became all the rage, in France and beyond.

The cravat gained ornamentation and flair throughout the seventeenth and eighteenth centuries, before scaling back in the nineteenth. But the necktie never went out of fashion. Often made of silk, there were a seemingly endless number of ways they could be tied, and the silk cravat was an essential part of any well-to-do (or aspiring well-to-do) man's wardrobe, for both business and pleasure.

Bowties and long ties began to appear in the late nineteenth century, and these erased most other options from the scene by the 1920s. After that, differences between neckties had more to do with shape and size, wide versus thin, short versus long, and so on. Ties are now made of silk, nylon, knits, and other fabrics, and appear in every color and pattern imaginable, from the beautiful to the bizarre. Love them or hate them, they are still considered a crucial part of looking sharp, and it's a safe bet they'll be hanging around a long time yet.

WHO MADE THE...
SHOELACE?

Most folks learn to tie their shoes when we're young, and let's be honest, it's a fairly big accomplishment! As it has been for most of human history. People have been tying shoes since at least 3500 BCE, and likely a lot longer than that.

As the concept and design are fairly simple, the shoe-lace was probably invented in more than one location simultaneously. But a lace only fufills its potential when threaded through holes in a piece of footwear, holding it tightly against a foot.

There might well be older surviving examples that archaeologists haven't yet unearthed, but at the moment, the oldest shoelace-like strings come from the so-called Areni-1 shoe, a one-piece leather shoe found at a cave site in 2008, in what is now southeastern Armenia. Its style is simple, drawing the leather up over the tops of the feet, where laces hold the fitting in place. The leather lace survives, as well, allowing researchers to get a bet-ter glimpse into shoe technology of the time.

It's thought that this kind of shoe was common not only in the area, but across a much larger region for centu-ries, at minumum. It's a simple, practical design that could have been replicated anywhere where people had access to animals and animal skins.

The next oldest shoe with laces dates from a few centuries later, ca. 3300 BCE. This belonged to the famous Ötzi, or "Iceman," a male body found in the Ötztal Alps on the border between Italy and Austria. The Iceman's shoes were made from a combination of bear and deer skin, suggesting that he and his people hunted the animals for their hides. Ötzi's footwear is more complex than the Armenian shoe, probably due to the desire to keep feet warm in the deep snow of the Alps. Importantly, the shoelaces are made of hemp and bark. This combination suggests an awareness of the need for strength in laces; after all, how annoying is it to be tying your shoe and have the lace break?

Laces for shoes, boots, and sandals have been considered essential ever since. No one knows who first came up with the idea, but it seems that cultures over a very large geographical region realized the need for them simultaneously.

WHO MADE...
COTTON CANDY?

It's a childhood favorite at county fairs, carnivals, circuses, and other grand spectacles. It is melt-in-your-mouth delicious, and looks way more substantial than it is. It's an indulgence featuring pure sugar and a bit of food coloring that is not one bit good for you, and that's part of what makes it so great! Cotton candy (aka candy floss in Great Britain) is machine-spun sugar that gets whipped up into a cotton-like texture and set on a stick. The idea is simple enough, and while it's credited with being invented at the end of the nineteenth century, its origins seem to extend much further back.

Some form of spun sugar was enjoyed as an upper-class dessert (sugar being very expensive back then) as long ago as the fifteenth century in Italy. This early form of the desert was made by hand, of course, and so couldn't have achieved anything like the fluffiness of modern spun sugar. But the groundwork of what would become cotton candy was in place.

Fast-forward to the year 1897, when an idea that was about to change fairs and carnivals forever arrived on the scene. A dentist named William Morrison got an idea for a machine that could spin candy. Yes, a dentist. Get the feeling that maybe he was looking for a way to increase his number of patients? Morrison was also a lawyer, so he must have had the legal angles figured out, too. In any case, he worked with a confectioner named John C. Wharton to produce such a machine. They perfected their design and exhibited it at the 1904 St. Louis World's Fair, where their "Fairy Floss" made a big splash. It's said that they made over $17,000 in sales from that one event!

Soon, other inventors wanted in on the idea, and the next year, Albert D. Robinson obtained his patent for an electric cotton candy machine, which he sold to General Electric two years later. By 1921, when the original patent had expired, yet another dentist named Josef Lascaux decided to get in on the action. Being from Louisiana, he thought the spun sugar (which is over 70 percent air) looked like the cotton that grew nearby, so he named his dessert "cotton candy," a name that has stuck ever since. Despite Lascaux's obvious marketing acumen, he was never especially successful in the candy business. Perhaps people had become a bit wary of this whole "dentists offering you sugar" thing.

WHO MADE...

NAIL POLISH?

Painted nails, both for fingers and toes, have a long history that stretches back centuries. And, like many fashions, they originally served to enforce class distinctions.

Painting nails is known to have been a practice in China at least 5,000 years ago, with polish being made from ingredients such as beeswax and egg whites. In ancient Chinese society, painted nails were a sign of wealth and status, and there were often restrictions on what colors the various classes could wear, red being reserved for the rich and nobility. If someone was caught wearing a color designated for those above their station, they could potentially be executed for it.

In Babylonian civilization, the warrior class was said to color their nails with kohl before going into battle, giving them that classic Goth look thousands of years before it became fashionable in the 1990s. Again, this color was strictly reserved for this class of individuals. Nail coloring was worn by women and men in Egypt, with henna serving as the preferred dye, providing the nails a reddish-brown look when it dried. There are mummies that have painted nails, showing how this bit of orna-mentation was so important that it was thought to be necessary for the afterlife.

Painting the nails with various substances went in and out of fashion (and fell in and out of official approval) for centuries, with upper-class folks frequently using substances and colors that were banned for those of lesser social status.

At some point in the 1870s, an American woman named May E. Cobb studied the French art of manicure (possibly in France itself; sources on the matter are a little unclear) and wanted to bring it back to her own country. In 1878, she opened "Mrs. Pray's Manicure," a salon in Manhattan. Being a very savvy businesswoman, Cobb soon expanded her business to other cities, such as Boston and Chicago. Eventually, she expanded into manufacturing her own line of cosmetics and nail-care products, including the emery board, which she and her husband invented.

The success of Cobb's cosmetic business caused others to hop on board, and in 1920 makeup artist Michelle Menard got the idea in 1920 of creating a glossy lacquer that reproduced the shine of automobile paint on the nails. In the 1930s, Menard's company, Revlon, introduced colors still in wide use today, the most popular being red. The love for this particular shade is thought to be tied to actress Rita Hayworth, who sparked a surge of interest while sporting it on silver screen.

WHO MADE...
EYEGLASSES?

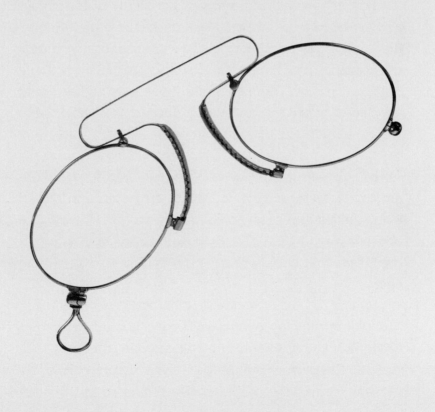

Many couldn't get around without them. With the number of different adverse conditions eyes are subject to, one wonders how people got along without eyeglasses before their advent. The answer: not too well. They either squinted their way to relative clarity, or lived in a miserable blur.

Inventors and thinkers as far back as the age of the Roman Empire realized that glass could be used to magnify or clarify an image. The mathematician Ptolemy mentioned this possibility in the second century, and by the year 1000, scholars in Europe and beyond were using "reading stones," hemisphere-shapes lenses that could be placed over text to make it larger; basically, these were rudimentary magnifying glasses.

In the thirteenth century, two English churchmen, Robert Grosseteste and Roger Bacon, wrote on the idea of using lenses to magnify objects and scripts. Clearly, the concept was in place. The leap needed to come in creating lenses that could enhance the abilities of the human eye.

The actual inventor of eyeglasses is, unfortunately, unknown, but we know they were being made in Pisa and Northern Italy by approximately 1290. These early versions came in pairs like modern glasses but did not have arms that hooked over the ears. The wearer had to hold them up to their face. In the sixteenth century, glasses were often held in place by hoops that looped over the ears. Because those who could read were the ones who most needed them, the wearing of glasses came to be associated with the intelligent and educated, a stereotype that persists.

These early glasses were basically magnification devices (like modern generic reading glasses), and the development of lenses for specific eye conditions was still centuries away. But they proved to be very popular among the literate and especially the elderly, whose sight was failing them in their older years. Plus, their convex lenses could be useful in correcting problems with reading text and seeing things at a distance. It wouldn't be until 1604 that German astronomer Johannes Kepler studied and wrote about how different lenses could correct various eye conditions. And of course, in the later eighteenth century, good old Benjamin Franklin would invent bifocals, giving people who had both myopia and farsightedness a chance to see everything much better! This opened up new territory in the nineteenth and twentieth centuries for lens design, frame design, and comfort, and innovations in eyeglasses technology are still ongoing.

WHO MADE...
CONTACT LENSES?

For many people, contact lenses are a miracle, allowing them to put away their cumbersome glasses and still see better than ever. For the squeamish, however, the very prospect of putting something directly on their eyeballs is a complete nonstarter. Contacts are an amazing invention for those who want them, but it might surprise you just how long the idea has been around. Having just taken a good look at eyeglasses (get it?), you might be surprised that contact lenses, at least the idea for them, came along just a few centuries later.

It was none other than Leonardo da Vinci who proposed the idea of using fitted lenses to alter one's vision. But his concept involved submerging one's head in a bowl of water, or placing a lens over the eye and pouring water into it. Not practical, not fun, not safe ... not a lot of things, really.

Other inventors took a stab at the idea over the centuries, but it wasn't until 1845 that English inventor and polymath Sir John Herschel proposed the idea of fitting a lens over the cornea, though Herschel doesn't seem to have tested his theory by creating a prototype. Other inventors took up the idea, but it was German ophthalmologist Adolf Gaston Eugen Fick who created the first successful corrective lens that could be placed upon the eye in 1888. The problem was, the lenses were heavy and covered over the whole eye, not just the cornea. Since our eyes receive their oxygen from the air, rather than from our blood, this meant that wearers were essentially choking their eyes, and the pain from wearing them became unbearable after an hour or two. Fick's effort was a bad initial design, but the concept was sound, and his model became the basis for the smaller contact lenses produced during the next several decades.

Indeed, the idea would be revisited in the 1930s, with the arrival of plastics. And in the late 1940s, English optical technician Kevin Touhy accidentally discovered that a smaller lens that only fit over the cornea would stay in place just as well, even when blinking.

In the late 1950s, Czechoslovakian chemist Otto Wichterle helped to create a new kind of plastic, hydrogel, which was softer and more pliable when wet, and could also be shaped and molded. Optometrist Dr. Robert Morrison heard about Wichterle's achievement and realized that a new kind of contact lens could be made from this material, one that would be much more comfortable to wear.

Lens manufacturers Bausch and Lomb realized the potential for this material as well and licensed it to start making their own contact lenses. They haven't looked back since. Contact lens technology continues to evolve and improve, making them even more comfortable than ever.

WHO MADE THE...
SWIMMING POOL?

Who doesn't love a dip in the pool on a hot summer day?
Of course, pools are also used for exercise and sporting
events, both indoors and outdoors. Places dedicated to
swimming other than natural bodies of water have been
with humanity for a very long time, even if initially they
were mainly just a luxury for the rich.

We don't know who first came up with the idea, but one
of the earliest man-made pools can be found at the Great
Baths of Mohenjo-Daro in what is now Pakistan, dating
from around 2600 BCE. This pool was made of bricks and
coated in plaster. Scholars and archaeologists are fairly
certain it wasn't intended for recreational use, however,
and probably served some religious function. Of course,
"religious purposes" is always the term used when
researchers can't figure out exactly what an ancient
object or site was used for. This place could have been
a vast kiddie pool for some nobleman's brood, for all we
know. Anyway, artificial pools served many purposes in
the ancient world, and the idea was widespread enough
to find its way into several cultures.

Both the Greeks and Romans built dedicated pools for
swimming (as opposed to bathing) to train young boys as
part of their education, athletes, and, perhaps, soldiers.
Roman politician Gaius Maecenas had the first known
heated pool installed at his home on the Esquiline Hill.

Gaius's luxurious spot was warmed by the central-heating system in his home.

Pools specifically for swimming went into a decline in Europe after the fall of the Western Roman Empire, though they were still enjoyed for bathing. Indeed, communal bathing pools for women and men were quite popular in the Middle Ages, despite being seen as places of licentiousness by the Catholic Church. A revival of interest in the swimming pool started in Britain in the 1820s and 1830s, when indoor pools were built in London and elsewhere. Indoor, because of the fickle British weather, of course! Swimming clubs and teams were formed, and notions of swimming gained popularity, so much so that it was designated as an Olympic sport for the first revived games in 1896. This made swimming even more popular, and more pools were built, in Britain and beyond.

The idea of swimming as sport and exercise remains, though for most, the pool simply remains the ultimate antidote to warm weather.

WHO MADE...
SUPER GLUE?

It's sticky enough to bond battleships together! And if you've ever gotten your fingers stuck to each other while using it, you know what a pain it is to remove! It's Super Glue, an invention that's very well-named. Although it has many uses today, its origins can be found in the 1940s, when it was being researched as a potential option for a very specific military purpose.

With World War II raging, chemist Harry Coover was working for the Eastman-Kodak company, experimenting with chemicals called cyanoacrylates to try to find a way to create reliable gunsights out of clear plastic. But Coover's trials were proving to be disappointing. Cyanoacrylates were far too sticky to serve the desired purpose, so Coover moved on.

Jumping ahead to 1951, Coover came to realize that cyanoacrylates might be very good for bonding things together, like battleships and wayward fingers. Actually, he and his colleagues were looking at jet canopies (i.e., the enclosure over a small jet's cockpit). Noticing that the stickiness of cyanoacrylates kept things together very well, he realized that this chemical compound could be a very effective glue for industrial purposes. He tested it out on multiple items, and sure enough, things stuck together, pretty much permanently.

Coover patented his discovery and Eastman-Kodak began looking for ways to market it. Having dubbed it "Super Glue," they found an enthusiastic response, as consumers seemed to be champing at the bit to stick stuff together. Coover even became something of a celebrity for his invention, appearing on television shows and in a commercial for the product.

One of the more creative and unexpected uses for Super Glue was discovered during the Vietnam War, when field doctors found that by spraying it on soldier's wounds they could close them up and stop the bleeding, allowing the wounded to be transferred safely back to a hospital. It was like stitching up a wound without having to stitch it, and it worked much more quickly. It's thought that Super Glue as a medical tool saved numerous lives, and the FDA would later approve its use in some surgical procedures.

Super Glue continues to be used today by industries and individuals alike. It has hundreds of uses, though it's still a pain if you get it on your fingers!

WHO MADE THE...
AIR
CONDITIONER?

It can be a literal lifesaver on a hot, humid day, and a metaphorical one for hospitals, data storage centers, and other essential businesses. Air conditioning allows people to live comfortably in brutally hot locations, though it's still way more common in the U.S. than in many other, much hotter countries. The idea of cooling the air for a living space is not at all new and has been employed around the world over the centuries.

Egypt, for example, gets very hot, and has been for a long time. The ancient Egyptians experimented with "passive cooling" devices, such as making use of shade to cool the surrounding air in a closed space (a courtyard with water and plants, for example) and then allowing that air to flow freely into home via open windows. There were various other techniques that operated on a similar principle, some of which worked better than others. These were simple ideas, but there was some merit to the idea of circulating cooled air though the house, even in these rudimentary forms.

Despite these efforts, for much of human history, people in very hot environments were just hot and miserable for a good portion of the year. However, during the nineteenth century, inventors began to think about methods to genuinely cool a building. In the 1840s, Dr. John Gorrie of Florida came up with an ice-making machine powered by horses. It never really took off, but it established the idea of artificial cooling as worthy of further study.

With increasing availability of electricity, air conditioning became much more viable. In 1902, inventor Willis Carrier came up with what is considered the first true air conditioner, which was installed in the offices of a publishing and lithograph company in Brooklyn. Willis would go on to form the Carrier Air Conditioning Company of America, which is still going strong today. In 1906, an engineer named Stuart W. Cramer devised a way to add moisture to the air at the textile mill where he worked. He combined this with the ventilation system to "condition" the air in the mill, and so called the process "air conditioning," a term that proved so sticky that Carrier also adopted it and included it in his company name.

This early air conditioning was meant for industry, and the idea of home air conditioning would have to wait until 1931, when H. H. Schultz and J. Q. Sherman invented the small air conditioner that could fit in a window. They were very expensive (well over $100,000 in modern money!), so their popularity was limited, to say the least. As with any invention, innovations to increase availability came along, and by the 1960s, small air conditioners were appearing in a lot of American homes.

More recently, air conditioning has been crucial for servers and computers, especially those connected to cryptocurrencies, a reality that has raised many questions and concerns about the long-term environmental effects of providing artificial coolness on a global scale.

WHO MADE...
LIQUID SOAP?

Bar soap was a game changer in terms of personal hygiene, but many people today prefer to use liquid soap, since it probably feels a bit "cleaner" than having to touch a bar of soap someone else's dirty hands have contacted. The viability of that claim can be debated, but there's no need to get worked into a lather wondering where liquid soap came from.

In 1865, William Shepphard was granted a patent for a liquid version of soap. Interestingly, the patent was for an "Improved Liquid Soap," which implies that an inferior liquid soap already existed. Or maybe it was just that the idea of soap itself was improved by putting it into a new form. It's possible that earlier forms of some kind of liquid soap were used in industrial settings, while Shepphard's was intended for personal hygiene within the home.

Then, in 1898, B. J. Johnson created soap made from palm oil and olive oil. He called it—you guessed it— "Palmolive." Despite these advances, liquid soaps were more often used for cleaning things than people. They were (and are) often concentrated detergents used to clean clothing and other surfaces. When it came time to clean oneself, most people remained firmly behind bars.

It wasn't until 1980 that the liquid soap as personal cleanser began to take off, when the Minnetonka Corporation saw an opening. One of the big issues with liquid soaps was the dispenser, or rather, the lack of one. They were either ineffective or you had to unscrew the bottle and pour some in your hand, like shampoo. Minnetonka felt that liquid soap needed a proper pump dispenser to really be appealing, and the company bought as many dispensers as it could (they were only made in a few factories at the time), in effect gaining a monopoly on them. Other companies, such the makers of Ivory and Irish Spring, were left out in the cold, at least for a few years.

This business gamble paid off, and the company's Soft Soap became a massive hit. Right off, people recognized the convenience of having a handy bottle that kept the soap in one place, as opposed to a slippery and potentially messy bar of soap. In 1987, Colgate-Palmolive bought Minnetonka, and the market for liquid soap exploded. Now liquid soaps are everywhere, and many people would never go back to a bar, for any reason. "Improved" indeed!

WHO MADE THE...
FLAT-SCREEN
TELEVISION?

Televisions have been the foundation of home entertainment for decades. The first models broadcast black-and-white images and were quite the luxury item. Nowadays, TVs come in an astonishing variety of designs and sizes, but the ones that everyone seems to want are the large, flat-screen models that allow people to re-create something of a movie-theater experience in their own homes. These screens can be modest to ridiculously large, depending on one's budget and the size of the room. They represent an amazing leap forward in technology from the old days of tube and solid-state TVs, with their chunky backs and hefty part–filled interiors.

So, who came up with this bright idea?

While the flat screen seems relatively new to consumers, engineers were already working on the idea as early as the 1950s. General Electric produced a prototype in 1954, but never pursued it for commercial use. Other companies came up with similar ideas, and in 1964, the first plasma-display, flat-panel screen was invented at the University of Illinois by professors Donald Bitzer and Gene Slottow, in response to a computer-rekated problem; it wasn't really their intention to make it into a commercially viable TV.

The flat screen took further steps in the 1960s with the development of liquid crystal displays (LCDs), a technology that found its way into conventional televisions. Finally, in 1996, Sony and Sharp decided to collaborate on developing a true flat-screen television for the general public. Well, not that general, because the first ones cost $15,000! But manufacturers knew they were onto something, and the possibility of a TV that could take up a lot less space and even be attached to a wall became increasingly appealing to consumers.

As interest and demand grew, innovation responded, making them easier and easier to produce, and prices plummeted. With advances in technology, the sizes of screens began to increase as well, from a max of approximately 30 inches or so initially to 103 inches at present. These outsized screens spurred the idea of a home theater, and these screens combined with the latest in sound technology and high-res video (such as Blu-Ray and modern high-def downloads) meant that viewers could have amazing cinematic experiences in their own homes.

Go into an electronics store today and you'll see nothing but a sea of flat screens of all sizes, beckoning to you. The seemingly overnight shift has been absolutely astonishing and shows how when an invention comes along at the right time, it can completely transform an industry and the lives of those who want it.

ABOUT THE AUTHOR

Tim Rayborn has written a large number of books and magazine articles (more than thirty each!), with a focus on music, the arts, general knowledge, and history. He will no doubt write more, whether anyone wants him to or not. He lived in England for many years and studied at the University of Leeds, which means he likes to pretend that he knows what he's talking about, but he's never actually invented anything.

He's also an almost-famous musician who plays dozens of unusual instruments from all over the world that most people of have never heard of and usually can't pronounce.

He has appeared on more than forty recordings, and his musical wanderings and tours have taken him across the US, all over Europe, to Canada and Australia, and to such romantic locations as Umbrian medieval towns, Marrakech, Vienna, Renaissance chateaux, medieval churches, and high school gymnasiums.

He currently lives among many books, recordings, and instruments, and a sometimes-demanding cat. He's pretty enthusiastic about good wines and cooking excellent food. Visit timrayborn.com for more on Tim and his work.

ABOUT CIDER MILL PRESS BOOK PUBLISHERS

Good ideas ripen with time. From seed to harvest, Cider Mill Press strives to bring fine reading, information, and entertainment together between the covers of its creatively crafted books. Our Cider Mill bears fruit twice a year, publishing a new crop of titles each spring and fall.

Visit us online at
cidermillpress.com

or write to us at
PO Box 454
12 Spring Street
Kennebunkport, Maine 04046